Micro/Nano Materials for Clean Energy and Environment

Micro/Nano Materials for Clean Energy and Environment

Special Issue Editors

Zhongchao Tan
Qinghai Li

MDPI • Basel • Beijing • Wuhan • Barcelona • Belgrade

MDPI

Special Issue Editors

Zhongchao Tan
University of Waterloo
Canada

Qinghai Li
Tsinghua University
China

Editorial Office
MDPI
St. Alban-Anlage 66
4052 Basel, Switzerland

This is a reprint of articles from the Special Issue published online in the open access journal *Materials* (ISSN 1996-1944) from 2018 to 2019 (available at: https://www.mdpi.com/journal/materials/special_issues/clean_energy)

For citation purposes, cite each article independently as indicated on the article page online and as indicated below:

LastName, A.A.; LastName, B.B.; LastName, C.C. Article Title. *Journal Name* **Year**, *Article Number*, Page Range.

ISBN 978-3-03921-128-9 (Pbk)
ISBN 978-3-03921-129-6 (PDF)

Cover image courtesy of Qinghai Li.

Contents

About the Special Issue Editors

Zhongchao Tan is a Professor in the Department of Mechanical and Mechatronics Engineering and Director of Research Lab for Green Energy and Pollution Control at the University of Waterloo. He is also the executive director of Tsinghua University-University of Waterloo Joint Research Center for Micro/Nano Energy & Environment Technology (https://uwaterloo.ca/jcmeet). He received his BSc and MSc degrees in Thermal Engineering from Tsinghua University, Beijing, China, in the 1990s, and PhD degree from the University of Illinois, Urbana Champaign, USA, in 2004. He then started his career as a faculty member at the University of Calgary, Canada from 2004 to 2010, and moved to the University of Waterloo in September 2010. His teaching and research are focused on filtration and separation technologies for green energy and clean environment.

Qinghai Li is an Associate Research Professor in the Department of Energy and Power Engineering at Tsinghua University, where he is also the director of Tsinghua University–University of Waterloo Joint Research Center for Micro/Nano Energy & Environment Technology (https://uwaterloo.ca/jcmeet). He received his doctoral, master, and bachelor degrees all from the Department of Thermal Engineering, Tsinghua University, China, with a minor degree in Environment Engineering. He worked as a visiting scholar in University of Sydney from February to August of 2016 and is a co-author of ca. 180 articles and several book chapters (http://rid.lib.tsinghua.edu.cn/scholar/652059). His research interests include photothermic/voltaic hybrid solar energy utilization, industrial boiler design, biomass, and waste combustion.

Preface to "Micro/Nano Materials for Clean Energy and Environment"

This Special Issue is dedicated to the first anniversary of the Tsinghua University–University of Waterloo Joint Research Center for Micro/Nano Energy and Environment Technology. I would like to extend my gratitude to the contributing scholars from the following institutes:

- Government College University, Pakistan;

- Guangdong Filtration and Wet Nonwoven Composite Materials Engineering Research Center, China;

- Institute of Archaeological Heritage—Monuments and Sites, Italy;

- SINOPEC, China;

- South China University of Technology, China;

- Taiyuan University of Technology, China;

- Tsinghua University, China;

- University of Minnesota, USA;

- University of Salento, Italy;

- University of Waterloo, Canada.

Eleven original works that describe recent advances in micro/nano materials in relation to clean energy and the environment are collected in this Special Issue. They are research papers from a broad range of topics related to micro/nanostructured materials aiming at future energy resources, low emission energy conversion, energy storage, energy efficiency, air emission control, air monitoring, air cleaning, and many other related applications. Energy and environment are two interrelated global challenges. We are confident that our scholars are contributing to a better international community.

Zhongchao Tan, Qinghai Li
Special Issue Editors

materials

MDPI

Article

Filtration of Sub-3.3 nm Tungsten Oxide Particles Using Nanofibrous Filters

Raheleh Givehchi [1], Qinghai Li [2,*] and Zhongchao Tan [1,2,*]

[1] Department of Mechanical & Mechatronics Engineering, University of Waterloo, Waterloo, ON N2L 3G1, Canada; raheleh.givehchi@utoronto.ca

[2] Tsinghua University—University of Waterloo Joint Research Centre for Micro/Nano Energy & Environment Technology, Tsinghua University, Beijing 100084, China

* Correspondence: liqh@tsinghua.edu.cn (Q.L.); tanz@uwaterloo.ca (Z.T.)

Received: 27 June 2018; Accepted: 23 July 2018; Published: 25 July 2018

Abstract: This work aims to understand the effects of particle concentration on the filtration of nanoparticles using nanofibrous filters. The filtration efficiencies of triple modal tungsten oxide (WO_x) nanoparticles were experimentally determined at three different concentrations for the size range of 0.82–3.3 nm in diameter. All tests were conducted using polyvinyl alcohol (PVA) nano-fibrous filters at an air relative humidity of 2.9%. Results showed that the filtration efficiencies of sub-3.3 nm nanoparticles depended on the upstream particle concentration. The lower the particle concentration was, the higher the filtration efficiency was.

Keywords: air filtration; airborne nanoparticle; particle concentration; nanofibers

---

1. Introduction

There has been a growing interest in filtration of airborne nanoparticles over the last decade primarily due to the concerns over the potential negative impact of nanoparticles on human health and the environment [1–3]. Filtration of nanoparticles is used in many applications such as respiratory protection, indoor air quality, and material synthesis. While the general mechanisms of nanoparticle filtration have been well understood, disagreements exist between experiments and theoretical analyses especially for nanoparticles approaching 1 nm in diameter due to their unique properties [4–9]. Heim et al. [10] showed that the filtration efficiency of sodium chloride (NaCl) nanoparticles in the range of 2.5–20 nm followed the classical single fiber-efficiency theory. Kim et al. [11] also tested the filtration efficiency of sub-100 nm particles and presented the independency of filtration efficiency on air humidity. They also showed that the classical filtration theory agrees well with the filtration efficiency of nanoparticles down to 2 nm. However, there is a deviation for sub-2 nm particles possibly due to a thermal rebound. Boskovic et al. [12] showed the dependency of filtration efficiency of nanoparticles on the particle shape for particles ranging from 50 nm to 300 nm at the velocity of 5 cm/s in which Brownian diffusion is dominant.

Several research groups have conducted a comprehensive literature review on nanoparticle filtration that led to different conclusions. Shaffer and Rengasamy [13] concluded that the conventional filtration theory can be used for the filtration performance of respirators for particles down to 4 nm in diameter. However, Mostofi et al. [14] concluded that one of the most challenging issues in nanofiltration is the lack of knowledge on the air flow rate, the temperature and humidity, and filter life. Wang and Otani [15] reviewed recent developments on nanoparticle filtration efficiency with a focus on the effect of thermal rebound, particle shape, aggregation, flow regime, air humidity, and particle loading. In addition, Wang and Tronville [16] summarized the advances in instrumentation for the filtration of nanoparticles down to 15 nm. Givehchi and Tan [4] provided an overview on studies of nanoparticle filtration and a thermal rebound. They concluded that little was known in this area of research.

While various studies have been completed on the filtration of airborne nanoparticles, limited data have been published to date on the effects of particle concentration on nanoparticle filtration efficiency. Understanding the behavior of nanoparticles down to 1 nm requires information on size distribution and composition of small nanoclusters [17]. According to conventional filtration theories, the filtration efficiency of particulate matter is independent of particle concentration. However, this was not validated for very small nanoparticles, which usually have a high number of concentrations. Ardkapan et al. [18] investigated the effect of particle concentration on the removal efficiency of an electrostatic fibrous filter. According to this research, nanoparticles up to 7 nm with a high concentration are captured with a higher rate due to the electrostatic forces rather than those with a low concentration. This knowledge is expected to be important for the design of high efficient filters, but it is missing in literature.

The objective of this study is to understand the effects of nanoparticle concentration on nanoparticle filtration. In the following section, the filtration efficiencies of WO_x particles ranging from 0.82 nm to 3.3 nm for triple modal number concentration distributions were experimentally determined. Experiments were conducted at three different particle concentrations using PVA nanofibrous filters at the relative humidity of about 2.9%.

2. Materials and Experimental Methods

2.1. Materials

Electrospun PVA nanofibrous filters made in a laboratory setting were used as the test filters. The characteristics of these nanofibrous filters have been described in other studies [19]. The solidity of filters is determined based on the measured pressure drop and filter thickness using the Davies equation [20]. Differences in the electrospinning parameters result in different mean fiber diameters, thicknesses, and solidities of the filters. Figure 1 shows the scanning electron microscope (SEM) images (Zeiss Gemini Model Leo 1550, Feldbach, Switzerland), which were taken at two magnifications, × 20K and × 5K, and fiber specifications of the sample.

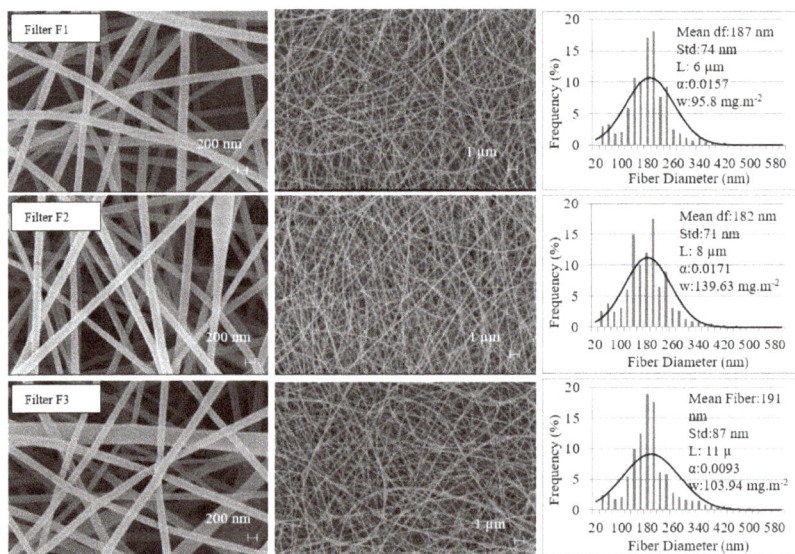

Figure 1. *Cont.*

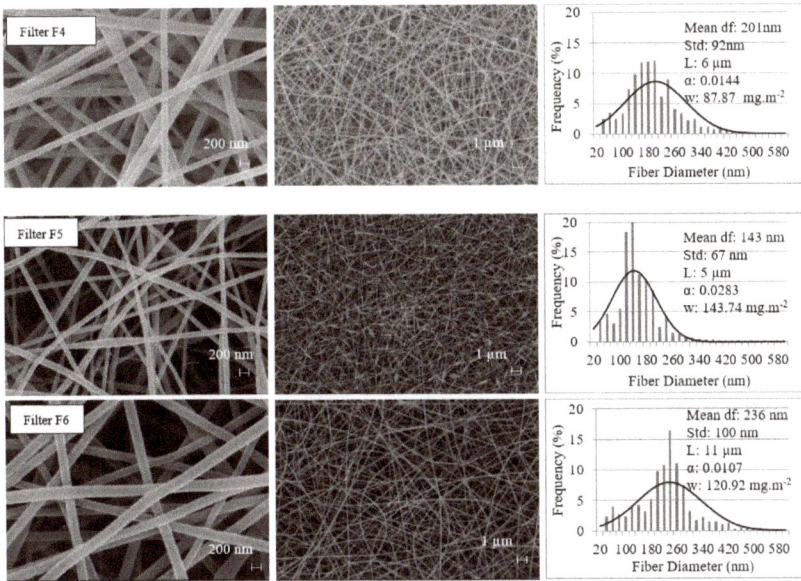

Figure 1. SEM images and fiber diameter distributions of nanofibrous filters.

2.2. Experimental Methods

As shown in Figure 2, the experimental setup used to determine the filtration efficiency of WO_x nanoparticles is similar to the one introduced in our previous publication [5] with an additional aerosol dilution system.

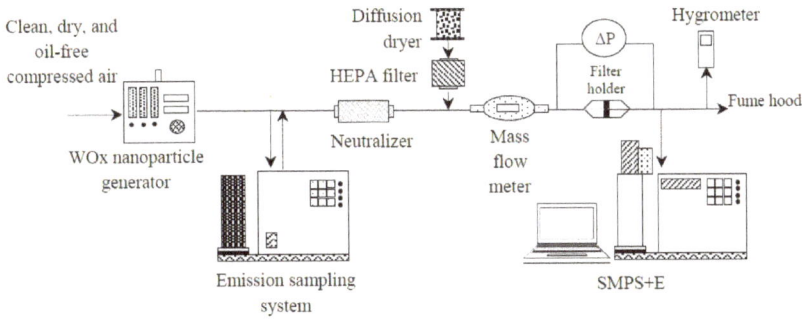

Figure 2. Schematic diagram of the experimental setup.

Nanoparticles down to 0.8 nm in diameter were generated using the WO_x nanoparticle generator (Model 7.860, GRIMM Aerosol techniK, Ainring, Germany). The size distribution remained stable during the experiments. The WO_x aerosol flow rate flow rate ranged from 7 $L \cdot h^{-1}$ to 10 $L \cdot h^{-1}$. The carrier air flow rate ranged from 120 $L \cdot h^{-1}$ to 200 $L \cdot h^{-1}$. The diluting air flow ranged from 200 $L \cdot h^{-1}$ to 250 $L \cdot h^{-1}$. Since the diffusion loss might be significant for such small particles, particle concentrations have to be large enough to be sufficient when the aerosol flow reaches the downstream measurement device. To detect these small particles, the aerosol flow rates were relatively high (4 lpm). A high flow rate is expected to shorten the travelling time of nanoparticles in the tubes and to reduce the diffusion loss of nanoparticles [21]. The relative humidity and temperature of the air flow during the experiments were 2.9 ± 0.4% and 24.7 ± 1.3 °C, respectively. This relative humidity does not represent real world nanoparticles. However, it was chosen because the particle generator generated nanoparticles at this small relative humidity. This low relative humidity is deemed to minimize the capillary force and minimize the adhesion energy between nanoparticles and filter fibers due to a capillary force [5].

The air emission sampling system (ESS, Model 7.917, GRIMM Aerosol techniK, Ainring, Germany) was employed to dilute the nano-aerosol for tests at lower concentrations. The dilution air and original aerosol were mixed in a counter flow mixer before passing through the aerosol cooler to reach room temperature. The dilution system is expected to prevent the condensation on nanoparticles and the formation of new nanoparticles. It also reduces the sample temperature to a level that is required by the measurement device. At the sample flow rate of 1 lpm, the dilution ratios were 1:10 and 1:100 for the first and second dilution stages, respectively. Note that particles travel in extra tubes due to the introduction of the ESS, which may further lower the particle concentration. Nanoparticles that survived were then passed through a radioactive neutralizer (Model P-2031, Staticmaster, Cincinnati, OH, USA), which was followed by the test filters.

The scanning mobility particle sizer coupled with a Faraday cup electrometer (SMPS+E, Model 5.706, GRIMM Aerosol techniK, Ainring, Germany) consists of a differential mobility analyzer (DMA, GRIMM Aerosol techniK). A Faraday cup electrometer (FCE, Model 5.705, GRIMM Aerosol techniK, Ainring, Germany) was used to measure nanoparticle size distribution before and after filtration. From previous studies, we are aware of the fact that many factors may lead to artifacts in the measured results. One of the most important issues associated with measuring particles in small sizes are high diffusion losses and low charging probability [22]. The resolution of DMA could be reduced by diffusing the broadening of small nanoparticles [23]. Some researchers related diffusion broadening to the observation of thermal rebound in relatively early publications [24–27]. Their experimental results were later challenged because of the accuracy of the DMA.

With this in mind, we attempted to minimize this kind of artifact. The first approach is to choose a Faraday cup electrometer over a condensation particle counter (CPC) as the particle detector downstream of the DMA. One critical challenge for CPC is that a high diffusional loss causes low counting efficiency for very small nanoparticles. The lower particle size detection limit of a commercial CPC is approximately 2 nm [28], which recently decreased to about 1 nm [29,30] and it is sensitive to the operating condition, the particle composition, and the charge state [29,31–33].

As an alternative device for detecting nanoparticles, FCE detects the charged nanoparticles at a response time of less than 100 ms [34] and it was believed to precisely detect nanoparticles of various compositions [28]. The lower detection limit of an FCE depends on the sensitivity of the specific FCE. GRIMM FCE was employed as a reference device for particles smaller than 3 nm [29] and FCE demonstrated a higher accuracy for smaller particles than CPC. Furthermore, FCE works well in high concentration samples [22]. For all these reasons, the GRIMM SMPS+E with short-DMA (S-DMA, GRIMM Aerosol techniK, Model 5.706) was employed in this research. It was capable of sizing and quantifying nanoparticles down to 0.8 nm. Furthermore, a sheath air to sample airflow ratio of 20:2 was used and remained constant in all experiments in order to minimize the diffusion loss of particles.

Both monodispersed and poly-dispersed nanoparticles have been used for filtration tests in literature. Some researchers employed monodispersed particles classified by a DMA [10,25,27,35–40]. Others used poly-dispersed particles and measured the particle number concentrations with an SMPS [41–53]. Possible errors when poly-dispersed particles are used can be eliminated by using a proper sampling method either by employing a time interval [48] or by introducing purge time [54] between measurements for an upstream and a downstream particle number concentration distribution. Both upstream and downstream concentrations reach equilibrium between consecutive sampling. Furthermore, a recent study confirmed that passing poly-dispersed nanoparticles and monodispersed nanoparticles through identical filters resulted in the same particle penetration measurement [55]. In the current study, therefore, the FCE and the DMA were set apart from each other at a minimum distance to reduce the nanoparticle loss due to diffusion. In our experiments, the particle size distribution measurements were stable as long as the particle concentrations were greater than 1000 particles/cm^3.

3. Results and Discussion

3.1. Filtration Efficiencies of Tested Filters for Sub-3.3 nm Particles

Figure 3 shows the size distribution of the WO$_x$ nanoparticles generated using the nano-aerosol generator. Each data point is averaged over at least three repeated measurements. The corresponding aerosol flow rate was 4 lpm. The tungsten air, the carrier air, and the diluting air flow rates in the tungsten aerosol generator were 10 L·h^{-1}, 220 L·h^{-1}, and 250 L·h^{-1}, respectively. The particle number concentrations of these particles remained stable.

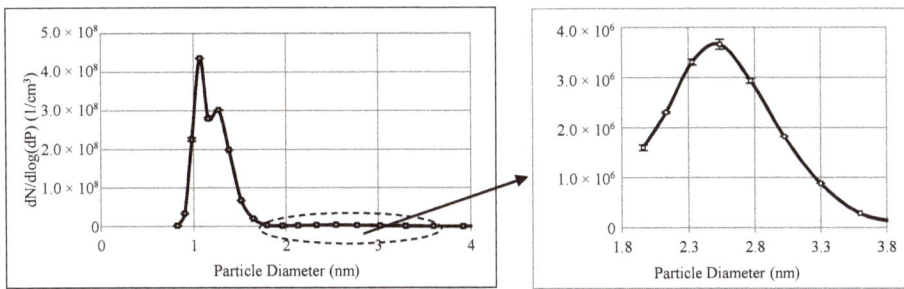

Figure 3. WO$_x$ nanoparticle size distribution generated by the tungsten oxide generator at an aerosol flow rate of 4 lpm.

As shown in Figure 3, the generated nanoparticles ranged from 0.82 nm to 4 nm in diameter. There are three peaks corresponding to 1.07 nm, 1.27 nm, and 2.54 nm. The particle concentrations for the first two peaks are in the order of 10^8 particles/cm^3 while the third peak is about 100 times lower than the other two (in the order of 10^6 particles/cm^3).

Figure 4 shows the upstream particle concentrations and the corresponding filtration efficiencies for six different electrospun nanofibrous filters. While the particle sizes ranged from 0.82 nm to 4 nm or more, the filtration efficiency was determined only for particles that ranged from 0.9 nm to 3.3 nm. This is because the particle concentrations of both lower and upper ends dropped below the lower detection limit of FCE in the air downstream of the filters. Comparing the filtration efficiencies for particles ranging from 0.9 nm to 3.3 nm for different tested filters shows that the measured filtration efficiencies depended on the concentrations of these nanoparticles. The measured filtration efficiencies for nanoparticles smaller than 1.96 nm are much lower than those for larger ones and there is a sharp drop in filtration efficiency when particle concentration increased as size decreased. Particle concentrations for sub-1.96 nm particles are in the range of 10^8 particles/cm^3. The concentrations of larger particles are around 10^6 particles/cm^3. The difference between the filtration efficiencies for these particles may be attributed to the differences in the incoming particle concentrations.

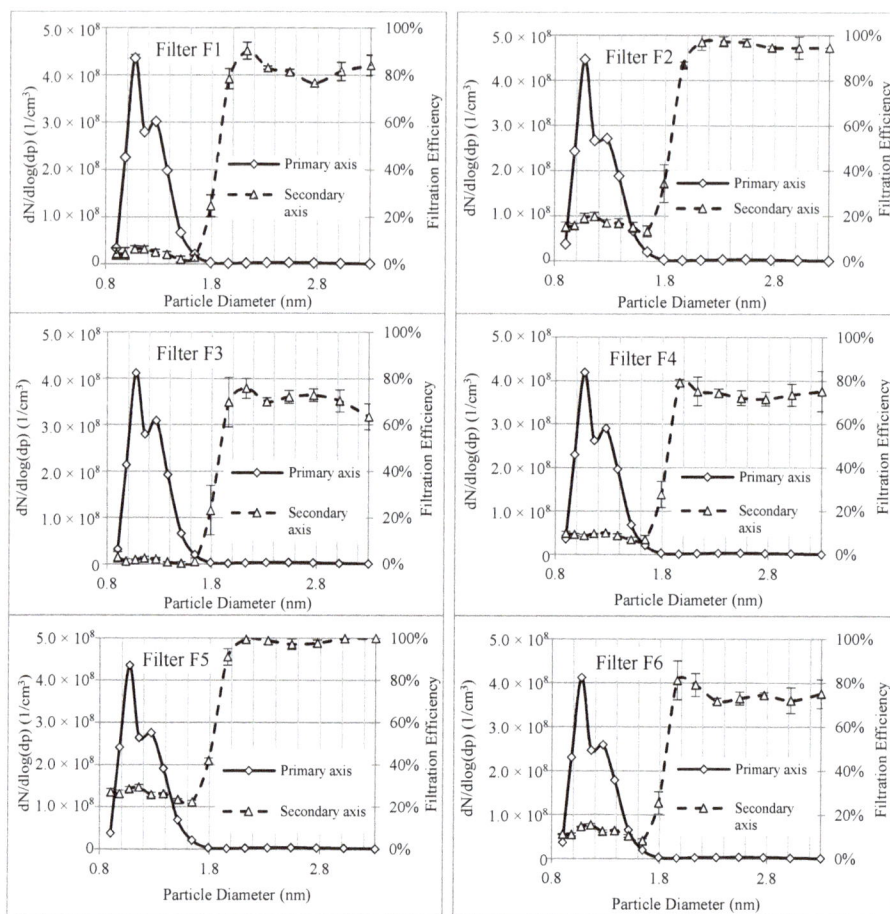

Figure 4. Nanoparticle size distributions along with particle removal efficiencies of different nanofibrous electrospun filters.

Further investigations show that the concentration of particles with a diameter of 1.79 nm is about the same as those with diameters of 2.77 nm and 2.33 nm. However, the filtration efficiency for 1.79 nm particles is much lower than the filtration efficiency of those with the other two diameters. In addition to particle concentrations, another mechanism may also contribute to the drop in filtration efficiency for smaller sized particles.

According to the conventional filtration model, the filtration efficiency increases as the size of small nanoparticles decreases [56]. However, results from this study showed clear drops in filtration efficiencies when the size of nanoparticles were below 1.96 nm. It appears that the conventional filtration model may need modification for these tested sizes of WO_x particles through the PVA nanofibrous filters and that the effect of particle concentration may have to be introduced into the models.

3.2. Effects of Particle Concentration on the Filtration of Nanoparticles

To systematically investigate the effects of nanoparticle concentration on air filtration, the ESS was employed to dilute the incoming aerosol prior to filtration tests. Since the original concentrations of particles larger than 1.96 nm in diameter were in the order of 10^6 particles/cm^3 (see Figure 3). The concentrations of large particles after dilution approached the detection limit of the FCE. Therefore, the effects of particle concentration on filtration efficiency are only presented in this study for sub-1.8 nm nanoparticles.

Three particle number concentration distributions, which correspond to no dilution, were diluted 10 times (1:10) and 100 times (1:100). These concentration distributions are shown in Figure 5. The particle concentrations have three different orders of magnitude from 10^8 to 10^7 and 10^6 particles/cm^3. The error bars in terms of standard deviation in the particle size range were less than 1% for a high particle concentration (in the order of 10^8), about 3% for the median (in the order of 10^7), and about 4% for a low particle concentration (in the order of 10^6). In all three figures, there are two peaks at 1.07 nm and 1.27 nm.

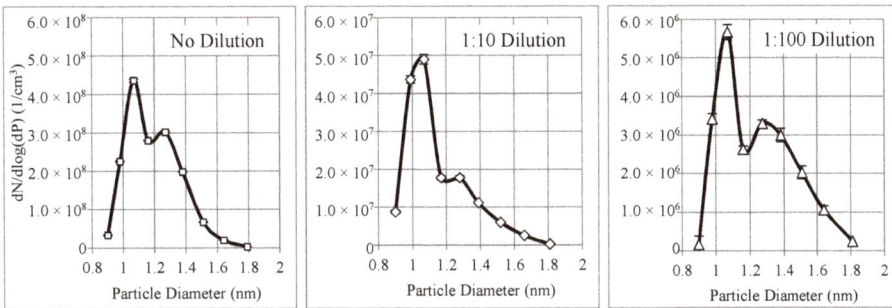

Figure 5. WO$_x$ nanoparticle number concentration distributions generated by the tungsten oxide generator at an aerosol flow rate of 4 lpm.

Figure 6 shows the measured filtration efficiencies of the electrospun nanofibrous filters for three cases from Figure 5. The results show clear dependencies of filtration efficiencies that rely on the incoming (upstream) particle concentration. The filtration efficiency increased with the dilution ratio. This observation is similar to the results reported by Shi and Ekberg (2015) except that the particle sizes were larger in their work. They also showed that the filtration efficiencies of particulate matter ranging from 300 nm to 500 nm decreased as the upstream particle concentration increased [57].

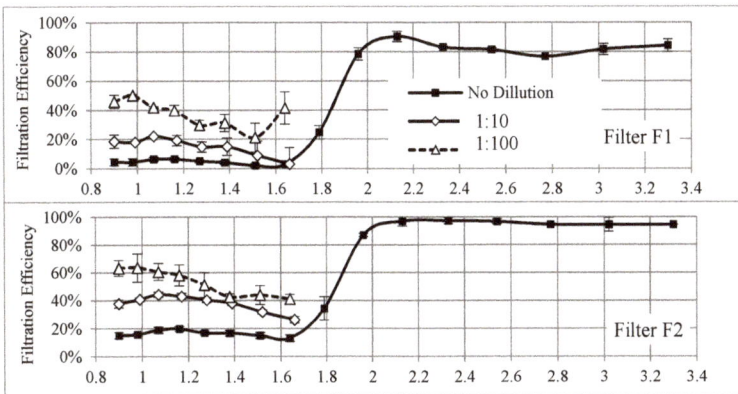

Figure 6. *Cont.*

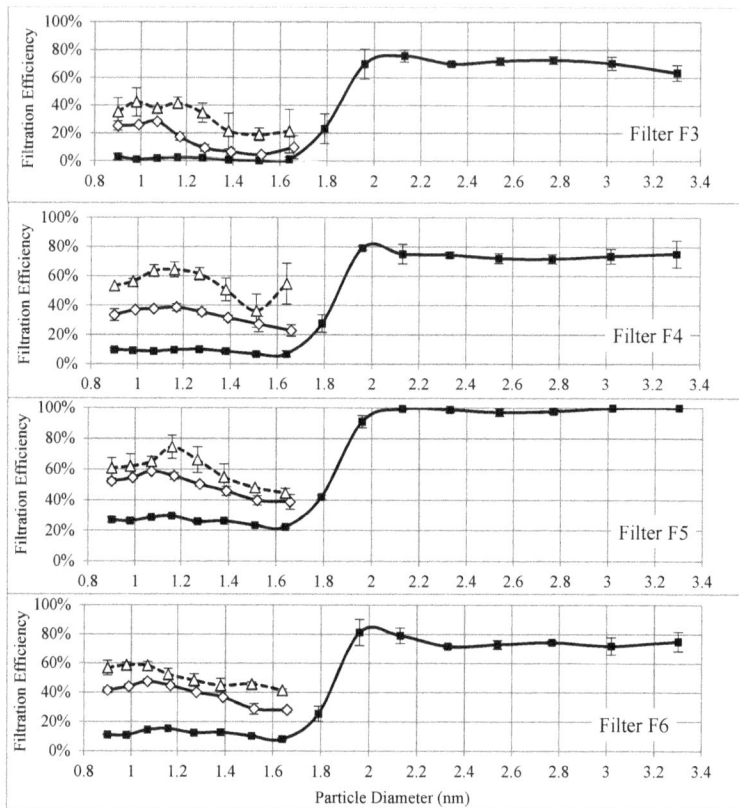

Figure 6. Particle filtration efficiencies of nanofibrous filters at three levels of particle concentrations (no dilution, 1:10 dilution, and 1:100 dilution).

The upstream particle concentrations for sub-1.8 nm particles when employing the second dilution stage of ESS are in the same magnitude as the concentrations in larger particles. Even at the same concentration level, the filtration efficiencies for sub-1.8 nm particles are lower than the filtration efficiencies of larger particles for all tested filters.

Figure 7 shows the filtration efficiencies for sub-1.8 nm particles as a function of upstream particle concentration for different nanofibrous filters. It was found that the measured filtration efficiencies also decreased with increasing particle concentration. Filter F5 showed the highest filtration efficiencies and Filter F3 showed the lowest filtration efficiencies for sub-1.8 nm particles at the three particle concentrations. Even though Filter F5 has the least thickness of 5 μm, it has the smallest mean fiber diameter (143 nm) and the greatest solidity (0.0283). The larger solidity to the mean fiber diameter ratio is likely to increase the filtration efficiency for nanoparticles at the price of a relatively large pressure drop of 228.93 Pa. Filter F3 has a relatively large thickness of 11 μm, which is thicker than other filters. Its mean fiber diameter is between those of Filters F1 and F4. However, it has the least solidity among all the filters, which decreases its filtration efficiency. The pressure drop of this filter is 42.30 Pa, which is lower than the pressure drop of other filters. Therefore, for these sizes of nanoparticles, which might behave like large gas molecules, a filter with a smaller mean fiber diameter and larger solidity has the highest filtration efficiency while the thickness of the filter has a negligible effect on the filtration efficiency among these nanoparticles.

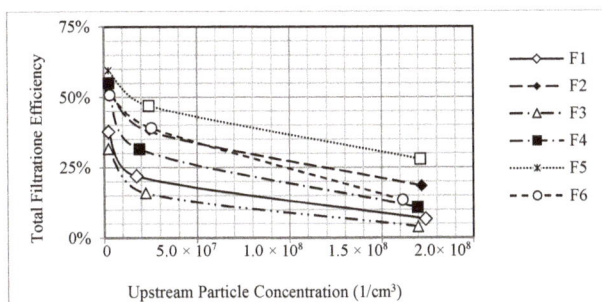

Figure 7. Total filtration efficiencies of electrospun filters as a function of upstream particle concentration.

3.3. Discussion on Thermal Rebound

Figure 6 also showed that the filtration efficiencies for particles smaller than 1.07 nm to 1.17 nm decreased as the size of nanoparticles decreased. This finding led us to revisit a possible thermal rebound that remains a debatable subject in literature. As explained above, nanofibrous filters and sub-1.8 nm WO_x particles were used in this study. In addition, all the filtration tests were conducted at a relatively low humidity. These parameters are different from other previous studies in thermal rebound. The majority of them could not experimentally prove the thermal rebound of nanoparticles. So far, only three studies have shown certain results of thermal rebound. Kim et al. [11] observed a drop in the filtration efficiency of a glass filter for uncharged sub-1.3 nm NaCl nanoparticles at a relative humidity of 1.22%. Van Gulijk and Schmidt-Ott [58] compared the penetration of different particles with a similar size distribution through a wire screen and found the possibility of thermal rebound for NaCl and $NiSO_4$ particles. However, they did not determine the particle critical diameter below which thermal rebound occurred. Rennecke and Weber [59] then investigated the thermal rebound of nanoparticles under low pressure and demonstrated the possibility of thermal rebound for dense NaCl particles ranging from 20 nm to 60 nm. They proposed that the thermal rebound was pressure dependent while, under normal conditions, gas friction may reduce the kinetic energy of a particle prior to rebounding and may cause its adhesion to a filter media surface.

We believe that the properties of filter media may also be important for the occurrence of thermal rebound. Most previous experimental studies employed commercial fibrous filters to test nanoparticle penetration through the filters, which led to conclusions that no thermal rebound was associated with their tested particles [37,47,52,54,60]. Thermal rebound may not be observed for thick and multilayer filters because rebounded particles can be captured by other fibers within the thick filter [4]. Consequently, single layer filters such as wire screens and thin nanofibrous filters are expected to minimize this artifact.

Two other earlier experimental studies showed the possibility of thermal rebound for sub-2 nm particles and they used wire screens [26,27]. However, wire screens have well—defined wires and openings. These open spaces may be sufficiently large for nanoparticles to pass through without collision on the wires. Therefore, the low efficiency of small nanoparticles may have been attributed inaccurately to a thermal rebound.

In the study, the tested nanofibrous filters are extra thin (L < 10 μm) and act like a single layer media. The media surface on which nanoparticles can collide is greatly increased due to the large surface area of nanofibers, which increases the possibility of thermal rebound if it does exist. If a particle rebounds from the nanofiber, there will be little chance for it to be captured again by other nanofibers. The filtration is then characterized by surface loading instead of depth loading because of the extra thin thickness of nanofibers.

Other important factors that may affect thermal rebound are the properties of nanoparticles. Although various methods have been used to produce test nano-aerosol for experimental studies of thermal rebound, very few can generate nanoparticles down to 1 nm with sufficiently high concentration. This was mainly limited by the availability of devices. Earlier studies that tested only nanoparticles greater than 2 nm did not show thermal rebound. Among a handful of papers where sub-2 nm particles were tested along with larger ones [11,25–27,30], three studies reported the possibility of thermal rebound for sub-2 nm particles [11,26,27]. Employing the sub-2 nm WO_x particles in the current study may be one of the factors that led to the observation of possible thermal rebound.

Several studies mentioned a charge effect as a reason why thermal rebound has not been observed yet. Particles are assumed to be neutral in the thermal rebound theory. However, charged particles induce an image force and reduce the rebound probability [30,58]. Heim et al. [30] investigated penetration of particles down to 1.2 onto wire grids. Based on this study, for the charged particles of WO_x and tetra-heptyl ammonium bromide, no thermal rebound was observed and the lower penetration of particles for a smaller particle is due to a small contribution by the image charge effect coupled with the diffusion. The generated nanoparticles in the current study are highly charged due to the effect of the thermal emission of electrons in the ceramic tube [30,61]. The majority of charged particles lost their charge in the neutralizer. However, a small fraction of charged particles remained and affected the results of thermal rebound.

The low relative humidity in this study may have also contributed to the observation of thermal rebound if it is the case. First, the adhesion energy between particles and filter media surfaces increased with the level of relative humidity due to the capillary force [5,62–65]. The increase in the adhesion energy may decrease the probability of thermal rebound.

The results herein showed that the critical diameters for thermal rebound were nearly the same for various dilution ratios. However, the drop-in filtration efficiencies (implying possible thermal rebound) is more obvious at lower particle concentrations (i.e., higher dilution ratios). This is also deemed reasonable. At lower particle concentrations, individual particles have more chances to collide with the nanofibers, which increases the probability of nanoparticle rebound from the surface and decreases the adhesion efficiency of particles to the surface.

There might be another hypothesis regarding the effect of particle concentration on thermal rebound. The high filtration efficiencies for nanoparticles at low concentrations likely resulted from the increased availability of the filtration surface area. This may be similar to the process of gas adsorption, which is concentration-dependent. Adsorption is a surface phenomenon caused by van der Waals forces [66] where gas molecules may stay on the surface of an adsorbent. In this process, as the concentration of gas molecules increases, more surfaces of solid are covered with gas molecules. Therefore, the availability of surface area decreases at higher concentrations [67]. It has been well accepted that nanoparticles behave differently than micron ones. Nanoparticles at these extremely small sizes (sub-1.8 nm) may behave like gas molecules or molecular clusters when they collide with the surface of the filter media. To quantify this concentration dependency, a mechanism that might be similar to adsorption can be considered for these extremely small nanoparticles.

All of the aforementioned factors may have led to a decreased rate of filtration efficiency for particles smaller than the critical diameters. The critical particle diameters are also almost constant for the differently tested nanofibrous filters. Since all tested filters were made of the same materials with the same mechanical constants (e.g., Hamaker constant, elastic mechanical constant, and specific adhesion energy), one would expect the particle critical diameter to be the same for the filtration of WO_x particles using PVA filters.

It would be useful to employ thin fibrous filters with micron scale fibers and determine if the same phenomenon occurs. Micron-scale fibers have a lower surface area to volume ratio than nano-scale fibers and it is expected that particles would cover more surfaces. Therefore, based on the initiative model proposed above, one would expect to observe similar trends for filters made of micron fibers. Furthermore, a systematic investigation by various types of particles and filters would lead to a better understanding in this subject. Additionally, considering the effect of particle bouncing and resuspension may also improve the results [68].

4. Conclusions

To summarize, this study showed that sub-3.3 nm WO_x particles smaller and larger than 1.96 nm behaved differently in air filtration. Filtration efficiencies dropped for particles with high particle concentrations. This study provides evidence of the existence of thermal rebound. For particles ranging from 1.07 nm to 1.17 nm, the reduction in filtration efficiency may be a result of thermal rebound. This reduction is more clear at lower particle concentrations because nanoparticles have more chance to collide with the surface of filter media.

Author Contributions: Conceptualization, Z.T. and Q.L.; Methodology, R.G.; Software, R.G.; Validation, R.G., Z.T. and Q.L.; Formal Analysis, R.G.; Investigation, R.G.; Resources, Q.L. and Z.T.; Data Curation, All; Writing-Original Draft Preparation, R.G.; Writing-Review & Editing, Z.T. and Q.L.; Visualization, R.G.; Supervision, Z.T.; Project Administration, Z.T.; Funding Acquisition, Z.T. and Q.L.

Funding: This study was funded by Natural Sciences and Engineering Research Council, Discovery Grant 31177-2010 RGPIN, the Canada Foundation for Innovation (CFI Grant 056215), and the National Key R&D Program of China (Grant No. 2017YFB0603901).

Conflicts of Interest: Authors declared no conflict of interest.

References

1. Castranova, V. The Nanotoxicology Research Program in NIOSH. *J. Nanopart. Res.* **2009**, *11*, 5–13. [CrossRef]
2. Ferreira, A.; Cemlyn-Jones, J.; Cordeiro, C.R. Nanoparticles, nanotechnology and pulmonary nanotoxicology. *Rev. Port. Pneumol.* **2013**, *19*, 28–37. [CrossRef] [PubMed]
3. Kreyling, W.G.; Semmler-Behnke, M.; Möller, W. Ultrafine particle-lung interactions: Does size matter? *J. Aerosol Med.* **2006**, *19*, 74–83. [CrossRef] [PubMed]
4. Givehchi, R.; Tan, Z. An Overview of Airborne Nanoparticle Filtration and Thermal Rebound Theory. *Aerosol Air Qual. Res.* **2014**, *14*, 45–63. [CrossRef]
5. Givehchi, R.; Tan, Z. The effect of capillary force on airborne nanoparticle filtration. *J. Aerosol Sci.* **2015**, *83*, 12–24. [CrossRef]
6. Wang, J. Effects of particle size and morphology on filtration of airborne nanoparticles. *KONA Powder Part. J.* **2012**, *30*, 256–266. [CrossRef]
7. Hosseini, S.A.; Tafreshi, H.V. On the importance of fibers' cross-sectional shape for air filters operating in the slip flow regime. *Powder Technol.* **2011**, *212*, 425–431. [CrossRef]
8. Givehchi, R.; Li, Q.; Tan, Z. The effect of electrostatic forces on filtration efficiency of granular filters. *Powder Technol.* **2015**, *277*, 135–140. [CrossRef]
9. Wang, Q.; Lin, X.; Chen, D. Effect of dust loading rate on the loading characteristics of high efficiency filter media. *Powder Technol.* **2016**, *287*, 20–28. [CrossRef]
10. Heim, M.; Mullins, B.J.; Wild, M.; Meyer, J.; Kasper, G. Filtration efficiency of aerosol particles below 20 nanometers. *Aerosol Sci. Technol.* **2005**, *39*, 782–789. [CrossRef]
11. Kim, C.S.; Bao, L.; Okuyama, K.; Shimada, M.; Niinuma, H. Filtration efficiency of a fibrous filter for nanoparticles. *J. Nanopart. Res.* **2006**, *8*, 215–221. [CrossRef]
12. Boskovic, L.; Agranovski, I.E.; Altman, I.S.; Braddock, R.D. Filter efficiency as a function of nanoparticle velocity and shape. *J. Aerosol Sci.* **2008**, *39*, 635–644. [CrossRef]
13. Shaffer, R.E.; Rengasamy, S. Respiratory protection against airborne nanoparticles: A review. *J. Nanopart. Res.* **2009**, *11*, 1661–1672. [CrossRef]
14. Mostofi, R.; Wang, B.; Haghighat, F.; Bahloul, A.; Jaime, L. Performance of mechanical filters and respirators for capturing nanoparticles—Limitations and future direction. *Ind. Health* **2010**, *48*, 296–304. [CrossRef] [PubMed]
15. Wang, C.; Otani, Y. Removal of nanoparticles from gas streams by fibrous filters: A review. *Ind. Eng. Chem. Res.* **2013**, *52*, 5–17. [CrossRef]

16. Wang, J.; Tronville, P. Toward standardized test methods to determine the effectiveness of filtration media against airborne nanoparticles. *J. Nanopart. Res.* **2014**, *16*, 2417. [CrossRef]

17. Junninen, H.; Ehn, M.; Petäjä, T.; Luosujärvi, L.; Kotiaho, T.; Kostiainen, R.; Rohner, U.; Gonin, M.; Fuhrer, K.; Kulmala, M. A high-resolution mass spectrometer to measure atmospheric ion composition. *Atmos. Meas. Tech.* **2010**, *3*, 1039–1053. [CrossRef]

18. Ardkapan, S.R.; Johnson, M.S.; Yazdi, S.; Afshari, A.; Bergsøe, N.C. Filtration efficiency of an electrostatic fibrous filter: Studying filtration dependency on ultrafine particle exposure and composition. *J. Aerosol Sci.* **2014**, *72*, 14–20. [CrossRef]

19. Givehchi, R.; Li, Q.; Tan, Z. Quality factors of PVA nanofibrous filters for airborne particles in the size range of 10–125 nm. *Fuel* **2016**, *181*, 1273–1280. [CrossRef]

20. Davies, C. The seperation of airborne dust and mist particles. *Proc. Inst. Mech. Eng.* **1952**, *1*, 185–213. [CrossRef]

21. Erickson, K.; Singh, M.; Osmondson, B. *Measuring Nanoparticle Size Distributions in Real-Time: Key Factors for Accuracy*; NSTI-Nanotech: Danville, CA, USA, 2007.

22. Kangasluoma, J.; Attoui, M.; Junninen, H.; Lehtipalo, K.; Samodurov, A.; Korhonen, F.; Sarnela, N.; Schmidt-Ott, A.; Worsnop, D.; Kulmala, M. Sizing of neutral sub 3nm tungsten oxide clusters using Airmodus Particle Size Magnifier. *J. Aerosol Sci.* **2015**, *87*, 53–62. [CrossRef]

23. Kousaka, Y.; Okuyama, K.; Adachi, M.; Mimura, T. Effect of Brownian diffusion on electrical classification of ultrafine aerosol particles in differential mobility analyzer. *J. Chem. Eng. Jpn.* **1986**, *19*, 401–407. [CrossRef]

24. Heim, M.; Mullins, B.J.; Kasper, G. Comment on: Penetration of ultrafine particles and ion clusters through wire screens by ichitsubo et al. *Aerosol Sci. Technol.* **2006**, *40*, 144–145. [CrossRef]

25. Alonso, M.; Kousaka, Y.; Hashimoto, T.; Hashimoto, N. Penetration of nanometer-sized aerosol particles through wire screen and laminar flow tube. *Aerosol Sci. Technol.* **1997**, *27*, 471–480. [CrossRef]

26. Otani, Y.; Emi, H.; Cho, S.J.; Namiki, N. Generation of nanometer size particles and their removal from air. *Adv. Powder Technol.* **1995**, *6*, 271–281. [CrossRef]

27. Ichitsubo, H.; Hashimoto, T.; Alonso, M.; Kousaka, Y. Penetration of ultrafine particles and ion clusters through wire screens. *Aerosol Sci. Technol.* **1996**, *24*, 119–127. [CrossRef]

28. Ankilov, A.; Baklanov, A.; Colhoun, M.; Enderle, K.; Gras, J.; Julanov, Y.; Kaller, D.; Lindner, A.; Lushnikov, A.A.; Mavliev, R.; et al. Particle size dependent response of aerosol counters. *Atmos. Res.* **2002**, *62*, 209–237. [CrossRef]

29. Wimmer, D.; Lehtipalo, K.; Franchin, A.; Kangasluoma, J.; Kreissl, F.; Kürten, A.; Kupc, A.; Metzger, A.; Mikkilä, J.; Petäjä, T. Performance of diethylene glycol-based particle counters in the sub-3 nm size range. *Atmos. Meas. Tech.* **2013**, *6*, 1793–1804. [CrossRef]

30. Heim, M.; Attoui, M.; Kasper, G. The efficiency of diffusional particle collection onto wire grids in the mobility equivalent size range of 1.2–8 nm. *J. Aerosol Sci.* **2010**, *41*, 207–222. [CrossRef]

31. Sem, G.J. Design and performance characteristics of three continuous-flow condensation particle counters: A summary. *Atmos. Res.* **2002**, *62*, 267–294. [CrossRef]

32. Kangasluoma, J.; Junninen, H.; Lehtipalo, K.; Mikkilä, J.; Vanhanen, J.; Attoui, M.; Sipilä, M.; Worsnop, D.; Kulmala, M.; Petäjä, T. Remarks on ion generation for CPC detection efficiency studies in sub-3-nm size range. *Aerosol Sci. Technol.* **2013**, *47*, 556–563. [CrossRef]

33. Kulmala, M.; Kontkanen, J.; Junninen, H.; Lehtipalo, K.; Manninen, H.E.; Nieminen, T.; Petaja, T.; Sipila, M.; Schobesberger, S.; Rantala, P.; et al. Direct observations of atmospheric aerosol nucleation. *Science* **2013**, *339*, 943–946. [CrossRef] [PubMed]

34. Keck, L.; Spielvogel, J.; Grimm, H. From nanoparticles to large aerosols: Ultrafast measurement methods for size and concentration. In Proceedings of the Nanosafe 2008: International Conference on Safe Production and Use of Nanomaterials, Grenoble, France, 3–7 November 2008.

35. Boskovic, L.; Altman, I.S.; Agranovski, I.E.; Braddock, R.D.; Myojo, T.; Choi, M. Influence of particle shape on filtration processes. *Aerosol Sci. Technol.* **2005**, *39*, 1184–1190. [CrossRef]

36. Chen, S.; Wang, J.; Fissan, H.; Pui, D.Y.H. Use of Nuclepore filters for ambient and workplace nanoparticle exposure assessment—Spherical particles. *Atmos. Environ.* **2013**, *77*, 385–393. [CrossRef]

37. Kim, S.C.; Harrington, M.S.; Pui, D.Y.H. Experimental study of nanoparticles penetration through commercial filter media. *J. Nanopart. Res.* **2007**, *9*, 117–125. [CrossRef]

38. Yang, S.; Lee, G.W.M. Filtration characteristics of a fibrous filter pretreated with anionic surfactants for monodisperse solid aerosols. *J. Aerosol Sci.* **2005**, *36*, 419–437. [CrossRef]
39. Thomas, D.; Mouret, G.; Cadavid-Rodriguez, M.C.; Chazelet, S.; Bémer, D. An improved model for the penetration of charged and neutral aerosols in the 4 to 80 nm range through stainless steel and dielectric meshes. *J. Aerosol Sci.* **2013**, *57*, 32–44. [CrossRef]
40. Furuuchi, M.; Eryu, K.; Nagura, M.; Hata, M.; Kato, T.; Tajima, N.; Sekiguchi, K.; Ehara, K.; Seto, T.; Otani, Y. Development and performance evaluation of air sampler with inertial filter for nanoparticle sampling. *Aerosol Air Qual. Res.* **2010**, *10*, 185–192. [CrossRef]
41. Balazy, A.; Toivola, M.; Reponen, T.; Podgorski, A.; Zimmer, A.; Grinshpun, S.A. Manikin-based performance evaluation of N95 filtering-facepiece respirators challenged with nanoparticles. *Ann. Occup. Hyg.* **2006**, *50*, 259–269. [PubMed]
42. Bałazy, A.; Toivola, M.; Adhikari, A.; Sivasubramani, S.K.; Reponen, T.; Grinshpun, S.A. Do N95 respirators provide 95% protection level against airborne viruses, and how adequate are surgical masks? *Am. J. Infect. Control* **2006**, *34*, 51–57. [CrossRef] [PubMed]
43. Brochot, C.; Mouret, G.; Michielsen, N.; Chazelet, S.; Thomas, D. Penetration of nanoparticles in 5 nm to 400 nm size range through two selected fibrous media. *J. Phys. Conf. Ser.* **2011**, *304*, 012068. [CrossRef]
44. Buha, J.; Fissan, H.; Wang, J. Filtration behavior of silver nanoparticle agglomerates and effects of the agglomerate model in data analysis. *J. Nanopart. Res.* **2013**, *15*, 359–369. [CrossRef]
45. Cyrs, W.D.; Boysen, D.A.; Casuccio, G.; Lersch, T.; Peters, T.M. Nanoparticle collection efficiency of capillary pore membrane filters. *J. Aerosol Sci.* **2010**, *41*, 655–664. [CrossRef]
46. Eninger, R.M.; Honda, T.; Adhikari, A.; Heinonen-Tanski, H.; Reponen, T.; Grinshpun, S.A. Filter performance of N99 and N95 facepiece respirators against viruses and ultrafine particles. *Ann. Occup. Hyg.* **2008**, *52*, 385–396. [PubMed]
47. Golanski, L.; Guiot, A.; Rouillon, F.; Pocachard, J.; Tardif, F. Experimental evaluation of personal protection devices against graphite nanoaerosols: Fibrous filter media, masks, protective clothing, and gloves. *Hum. Exp. Toxicol.* **2009**, *28*, 353–359. [CrossRef] [PubMed]
48. Lore, M.B.; Sambol, A.R.; Japuntich, D.A.; Franklin, L.M.; Hinrichs, S.H. Inter-laboratory performance between two nanoparticle air filtration systems using scanning mobility particle analyzers. *J. Nanopart. Res.* **2011**, *13*, 1581–1591. [CrossRef]
49. Otani, Y.; Eryu, K.; Furuuchi, M.; Tajima, N.; Tekasakul, P. Inertial classification of nanoparticles with fibrous filters. *Aerosol Air Qual. Res.* **2007**, *7*, 343–352. [CrossRef]
50. Rengasamy, S.; Eimer, B.C.; Shaffer, R.E. Comparison of nanoparticle filtration performance of NIOSH-approved and CE-marked particulate filtering facepiece respirators. *Ann. Occup. Hyg.* **2009**, *53*, 117–128. [CrossRef] [PubMed]
51. Golshahi, L.; Abedi, J.; Tan, Z. Granular filtration for airborne particles: Correlation between experiments and models. *Can. J. Chem. Eng.* **2009**, *87*, 726–731. [CrossRef]
52. Huang, S.; Chen, C.; Chang, C.; Lai, C.; Chen, C. Penetration of 4.5 nm to 10 μm aerosol particles through fibrous filters. *J. Aerosol Sci.* **2007**, *38*, 719–727. [CrossRef]
53. Leskinen, J.; Joutsensaari, J.; Lyyränen, J.; Koivisto, J.; Ruusunen, J.; Järvelä, M.; Tuomi, T.; Hämeri, K.; Auvinen, A.; Jokiniemi, J. Comparison of nanoparticle measurement instruments for occupational health applications. *J. Nanopart. Res.* **2012**, *14*, 718. [CrossRef]
54. Japuntich, D.A.; Franklin, L.M.; Pui, D.Y.; Kuehn, T.H.; Kim, S.C.; Viner, A.S. A comparison of two nano-sized particle air filtration tests in the diameter range of 10 to 400 nanometers. *J. Nanopart. Res.* **2007**, *9*, 93–107. [CrossRef]
55. Li, L.; Zuo, Z.; Japuntich, D.A.; Pui, D.Y.H. Evaluation of filter media for particle number, surface area and mass penetrations. *Ann. Occup. Hyg.* **2012**, *56*, 581–594. [PubMed]
56. Hinds, W.C. *Aerosol Technology Properties, Behavior, and Measurement of Airborne Particles*; Wiley—Interscience: Hoboken, NJ, USA, 1999.
57. Shi, B.; Ekberg, L. Ionizer Assisted Air Filtration for Collection of Submicron and Ultrafine Particles—Evaluation of Long-Term Performance and Influencing Factors. *Environ. Sci. Technol.* **2015**, *49*, 6891–6898. [CrossRef] [PubMed]
58. Van Gulijk, C.; Bal, E.; Schmidt-Ott, A. Experimental evidence of reduced sticking of nanoparticles on a metal grid. *J. Aerosol Sci.* **2009**, *40*, 362–369. [CrossRef]

59. Rennecke, S.; Weber, A.P. On the pressure dependence of thermal rebound. *J. Aerosol Sci.* **2013**, *58*, 129–134. [CrossRef]

60. Van Osdell, D.W.; Liu, B.Y.H.; Rubow, K.L.; Pui, D.Y.H. Experimental study of sub-micrometer and ultrafine particle penetration and pressure drop for high efficiency filters. *Aerosol Sci. Technol.* **1990**, *12*, 911–925. [CrossRef]

61. Peineke, C.; Schmidt-Ott, A. Explanation of charged nanoparticle production from hot surfaces. *J. Aerosol Sci.* **2008**, *39*, 244–252. [CrossRef]

62. Bateman, A.P.; Belassein, H.; Martin, S.T. Impactor apparatus for the study of particle rebound: Relative humidity and capillary forces. *Aerosol Sci. Technol.* **2014**, *48*, 42–52. [CrossRef]

63. Stein, S.W.; Turpin, B.J.; Cai, X.; Huang, P.; Mcmurry, P.H. Measurements of relative humidity-dependent bounce and density for atmospheric particles using the DMA-impactor technique. *Atmos. Environ.* **1994**, *28*, 1739–1746. [CrossRef]

64. Chen, S.; Tsai, C.; Chen, H.; Huang, C.; Roam, G. The influence of relative humidity on nanoparticle concentration and particle mass distribution measurements by the MOUDI. *Aerosol Sci. Technol.* **2011**, *45*, 596–603. [CrossRef]

65. Pakarinen, O.; Foster, A.; Paajanen, M.; Kalinainen, T.; Katainen, J.; Makkonen, I.; Lahtinen, J.; Nieminen, R. Towards an accurate description of the capillary force in nanoparticle-surface interactions. *Model. Simul. Mater. Sci. Eng.* **2005**, *13*, 1175. [CrossRef]

66. Tan, Z. *Air Pollution and Greenhouse Gases: From Basic Concepts to Engineering Applications for Air Emission Control*; Springer: Berlin, Germany, 2014.

67. Namasivayam, C.; Muniasamy, N.; Gayatri, K.; Rani, M.; Ranganathan, K. Removal of dyes from aqueous solutions by cellulosic waste orange peel. *Bioresour. Technol.* **1996**, *57*, 37–43. [CrossRef]

68. Boor, B.E.; Siegel, J.A.; Novoselac, A. Monolayer and multilayer particle deposits on hard surfaces: Literature review and implications for particle resuspension in the indoor environment. *Aerosol Sci. Technol.* **2013**, *47*, 831–847. [CrossRef]

materials

MDPI

Article

Theoretical Study of As$_2$O$_3$ Adsorption Mechanisms on CaO surface

Yaming Fan [1,2,†], Qiyu Weng [2,3,4,†], Yuqun Zhuo [2,3,4,*], Songtao Dong [1], Pengbo Hu [2,3,4] and Duanle Li [2,3,4]

[1] Research Institute of Petroleum Processing, SINOPEC, Beijing 100083, China; fanymthu@163.com (Y.F.); dongst.ripp@sinopec.com (S.D.)
[2] Key Laboratory for Thermal Science and Power Engineering of the Ministry of Education, Department of Energy and Power Engineering, Tsinghua University, Beijing 100084, China; wqy17@mails.tsinghua.edu.cn (Q.W.); hupb18@mails.tsinghua.edu.cn (P.H.); liduanle@163.com (D.L.)
[3] Tsinghua University-University of Waterloo Joint Research Center for Micro/Nano Energy and Environment Technology, Tsinghua University, Beijing 100084, China
[4] Beijing Engineering Research Center for Ecological Restoration and Carbon Fixation of Saline–alkaline and Desert Land, Tsinghua University, Beijing 100084, China
* Correspondence: zhuoyq@mail.tsinghua.edu.cn
† These authors contributed equally to this work.

Received: 23 January 2019; Accepted: 18 February 2019; Published: 25 February 2019

Abstract: Emission of hazardous trace elements, especially arsenic from fossil fuel combustion, have become a major concern. Under an oxidizing atmosphere, most of the arsenic converts to gaseous As$_2$O$_3$. CaO has been proven effective in capturing As$_2$O$_3$. In this study, the mechanisms of As$_2$O$_3$ adsorption on CaO surface under O$_2$ atmosphere were investigated by density functional theory (DFT) calculation. Stable physisorption and chemisorption structures and related reaction paths are determined; arsenite (AsO$_3$$^{3-}$) is proven to be the form of adsorption products. Under the O$_2$ atmosphere, the adsorption product is arsenate (AsO$_4$$^{3-}$), while tricalcium orthoarsenate (Ca$_3$As$_2$O$_8$) and dicalcium pyroarsenate (Ca$_2$As$_2$O$_7$) are formed according to different adsorption structures.

Keywords: CaO; As$_2$O$_3$; DFT; adsorption

1. Introduction

Arsenic is a hazardous element existing in fossil fuels such as coal and petroleum [1]. According to the properties of arsenic and its compounds, it has been classified as volatile trace element by Clark and Sloss [2]. During combustion or chemical industry processes, gaseous arsenic is released into the environment. Excess amounts of arsenic can pollute water and soil. The exposure of arsenic to human may lead to hyperpigmentation, keratosis, skin and lung cancers with high possibility [3,4]. Arsenic compounds (including inorganic arsine) have been identified as hazardous air pollutants by the US government since 1990 [5]. The concentration of atmospheric arsenic in China is 51.0 ng/m^3, which is much higher than the limit of NAAQS (6.0 ng/m^3, GB 3095-2012) and the limit of WHO (6.6 ng/m^3, WHO) [6].

Combustion of fossil fuels, especially coal, is one of the main sources for anthropogenic emission of atmospheric arsenic [7]. It was estimated that 335.5 tons of atmospheric arsenic were emitted from Chinese coal-fired plants in 2010 [8]. In 2011, the US Environmental Protection Agency issued the Mercury and Air Toxics Standards (US, MATS, updated in 2016). An arsenic emission limit of 3.0×10^{-3} lb/MWh (approximately 0.41 µg/m^3) was set for coal-fired power plants [9]. In Chinese coal-fired power plants, the control of arsenic still remains scarce, but there are increasing interests in understanding its transformation in flue gas and developing emission reduction techniques.

Under an oxidizing atmosphere, gaseous As_2O_3 should be the main form of arsenic combustion products [10]. It has been proven that CaO could adsorb As_2O_3 in the coal-fired flue gas, and the dominating products were arsenate (AsO_4^{3-}) [11–14]. CaO component in fly ash leads to the enrichment of arsenic [15–18]. R.O. Sterling [11] found that CaO could effectively adsorb As_2O_3 at 600 °C and 1000 °C; the adsorption products were $Ca_3As_2O_8$ when O_2 existed. Jadhav [12] studied the adsorption products of As_2O_3 on a CaO surface under O_2 atmosphere between 300 °C and 1000 °C. X-ray photoelectron spectroscopy (XPS) and X-ray Diffraction (XRD) reflected that, when temperature was lower than 600 °C, the adsorption product was $Ca_3As_2O_8$; when temperature was between 700 °C and 900 °C, the adsorption products was $Ca_2As_2O_7$; and when temperature was as high as 1000 °C, the adsorption product was $Ca_3As_2O_8$. He also revealed that SO_2 and HCl played a weak role in adsorption. Li [13,14] studied the influence of CO_2 and SO_2 on the capture of As_2O_3 by CaO. The existence of SO_2 and CO_2 did not change the form of arsenic in adsorption products. The previous study certified the strong adsorption of As_2O_3 on CaO surface and the important role O_2 played in the reaction. However, the acute toxicity and low concentration of arsenic significantly limit the experimental research of As_2O_3 adsorption. The adsorption mechanisms still remain unclear, especially the composition of adsorption active sites and product structures.

Quantum chemistry calculation based on density functional theory (DFT) has become an effective method to simulate structures [19] and surface reaction of volatile trace elements [20]. For example, the adsorption of As^0 on a CaO (001) surface has been effectively studied by Zhang [21]. In this study, the adsorption structures and the detailed adsorption steps between the CaO surface and As_2O_3 (under O_2 atmosphere) have been studied by advanced DFT calculation, with the aim to offer microscopic information about critical reactions, and thus, to provide guidance to develop more efficient adsorbents and related control technologies.

2. Methods and Modeling

2.1. Methods

The material studio CASTEP [22,23] module was applied in the DFT calculation. The GGA (Generalized Gradient Approximation) and PBE [24] (Perdew-Burke-Ernzerhof) were chosen to describe the exchange and correlation interactions. The electronic wave functions were expanded on a plane wave basis with cut-off energy of 380 eV. The ultra-soft pseudo potential was referred to describe the interactions between electrons and the ionic cores [25]. 'The spin-polarized' option was selected for 'spin-unrestricted' calculations [26]. The BFGS (Broyden-Flechter-Goldfarb-Shanno) optimization algorithm was chosen for geometry optimization [27]. The transition state and reaction path (intermediate states) was determined by using the complete Linear Synchronous Transit/Quadratic Synchronous Transit (LST/QST) method [28] and confirmed by the Nudged-Elastic Band (NEB) method [29].

The convergence criteria of geometry optimization included: (a) self-consistent field (SCF) of 5.0×10^{-7} eV/atom; (b) energy of 5×10^{-6} eV/atom; (c) displacement of 5×10^{-4} Å; (d) force of 0.01 eV/Å; and (e) stress of 0.02 GPa. The convergence of complete LST/QST method (RMS, Root Mean Square) was set to 0.05 eV/Å. The convergence criteria of NEB included: (a) energy of 1.0×10^{-5} eV/atom; (b) max force of 0.05 eV/Å; and (c) max displacement of 0.004 Å.

The adsorption energy (E_{ads}) was defined as follows:

$$E_{ads} = E_{pro} - (E_{slab} + E_{adsorbate}) \tag{1}$$

where E_{pro} was the total energy of adsorption product, E_{slab} was the total energy of the slab model, and $E_{adsorbate}$ was the total energy of isolated adsorbate As_2O_3 or O_2 at its equilibrium geometry. A negative E_{ads} value represented a stable adsorption system.

2.2. Modeling

The energy of CaO crystal cell was converged with $6 \times 6 \times 6$ k points in the Monhorst-pack grid [30]. The equilibrium geometry of As_2O_3 and O_2 was examined in a cell of $20 \times 20 \times 20$ $Å^3$ periodic box. As shown in Table 1, the values of the calculated bond lengths, angles, and lattice parameters are consistent with the data reported from the previous study, indicating the reliability of the calculation.

Table 1. Calculated lattice parameters, bond lengths, and bond angles.

Substance	Previous Data	Simulated Data
CaO [31,32]	4.836 Å/4.807 Å	4.837 Å
As_2O_3 [33]	As–O bond 1.794 Å As–O bond 1.610 Å O–As–O angle 106.3° As–O–As angle 133.8°	As–O bond 1.814 Å As–O bond 1.622 Å O–As–O angle 111.2° As–O–As angle 141.8°
O_2 [34]	O–O 1.210 Å	O–O 1.240 Å

In our previous study, the CaO(001) slab model has been widely used for CO_2 [35], Se^0 [36] and SeO_2 [37] heterogeneous adsorption reaction, in which the good consistency with experimental work has been proven. Similarly, a 4-layer 3×3-surface CaO (001) slab was modeled to describe the CaO surface between CaO and As_2O_3 in this study. The superficial two layers of atoms were relaxed while the rest layers were fixed [38]. The vacuum region between slabs was set to 10 Å to avoid interactions among periodic images [39]. The energy of slab models and related adsorption structures were converged with $2 \times 2 \times 1$ k points in the Monhorst-pack grid. The detailed modeling process was put in the Supplementary Materials (Optimization of slab model section: Tables S1 and S2).

3. Results and Discussions

According to the spatial position of As_2O_3 and surface atoms distribution, three groups, including twenty-one possible As_2O_3 structures, were first modeled as the initial structures for optimization (provided in Figure S1). After the geometric optimization of the initial structures, plenty of adsorption structures were validated, then the possible reaction paths were calculated. Based on the minimal point of the reaction paths, additional stable structures were acquired. Most of the physisorption structures were similar in terms of structural pattern and close in terms of energy level; thus, three representative physisorption structures (adsorption energy higher than -100 kJ/mol [40]) were determined. Additionally, ten chemisorption structures (adsorption energy lower than -100 kJ/mol [40]) were identified. Based on these structures, various adsorption paths were finally confirmed. For briefness, the nth physisorption structure was abbreviated as P_n, while the nth chemisorption structure was abbreviated as C_n.

3.1. Stable Sorption Structures

3.1.1. Stable Physisorption Structures

Three representative physisorption structures have been shown in Table 2. The dominating differences are the number of As_2O_3's O bonded with superficial Ca and the distribution of the superficial Ca occupied by As_2O_3's O. Three types of physisorption follow the crystal orientation <100>, <110> and <110>, respectively. Two or three superficial Ca is close to As_2O_3's O, and the bond length is about 2.380 Å to 2.876 Å. The corresponding adsorption energy ranges from -65.8 kJ/mol to -58.4 kJ/mol. Based on electron density cloud, physisorption active sites are composed of superficial Ca atoms that interact with O of As_2O_3.

Table 2. Stable physisorption structures, adsorption energy, electron density cloud, and E_{ads}.

Name	Top View	Front View	Electron Density Cloud	Structure Details	E_{ads}
P_1				Bond$_{12}$: 2.450 Å Bond$_{34}$: 2.469 Å	−65.8 kJ/mol
P_2				Bond$_{12}$: 2.380 Å	−62.6 kJ/mol
P_3				Bond$_{12}$: 2.876 Å Bond$_{34}$: 2.539 Å	−58.4 kJ/mol

Ca(CaO)   O(CaO)   As(As$_2$O$_3$)   O(As$_2$O$_3$)

3.1.2. Stable Chemisorption Structures

Ten chemisorption structures were obtained, with E_{ads} ranging from −198.5 kJ/mol to −391.4 kJ/mol, which implies strong chemisorption. Superficial Ca is close to As$_2$O$_3$'s O, the bond length is about 2.269 Å to 2.528 Å, while superficial O is close to As$_2$O$_3$'s O, the bond length is 1.788 Å to 2.086 Å. According to electron density cloud and bong length, chemisorption active sites are superficial O atoms that interact with As of As$_2$O$_3$. According to the adsorption energy and structure (i.e., the positions of As and O), four categories were classified in Table 3:

Category I: As$_2$O$_3$'s As is located on the hollow site
Category II: All of As$_2$O$_3$'s O is located on or close to superficial Ca top site
Category III: As$_2$O$_3$ transforms into a spoon-shaped structure
Category IV: All of As$_2$O$_3$'s As is located on two neighboring superficial O top site

Table 3. Chemisorption structures, adsorption energy, electron density cloud and E_{ads}.

Category	Name	Top View	Front View	Electron Density Cloud	Structure Details	E_{ads}
I	C_1				Bond$_{12}$: 2.635 Å Bond$_{14}$: 2.086 Å Bond$_{35}$: 2.360 Å	−198.5 kJ/mol
II	C_2				Bond$_{12}$: 1.858 Å Bond$_{34}$: 2.360 Å Bond$_{56}$: 2.386 Å	−222.1 kJ/mol
II	C_3				Bond$_{12}$: 2.424 Å Bond$_{34}$: 1.815 Å Bond$_{56}$: 2.314 Å Bond$_{78}$: 2.490 Å	−274.4 kJ/mol

Table 3. *Cont.*

Category	Name	Top View	Front View	Electron Density Cloud	Structure Details	E_{ads}
II	C_4				Bond$_{12}$: 2.269 Å Bond$_{34}$: 1.949 Å Bond$_{56}$: 2.391 Å	-292.0 kJ/mol
II	C_5				Bond$_{12}$: 2.293 Å Bond$_{34}$: 1.943 Å Bond$_{56}$: 2.298 Å	-315.1 kJ/mol
III	C_6				Bond$_{12}$: 1.815 Å Bond$_{34}$: 2.355 Å	-302.3 kJ/mol
III	C_7				Bond$_{12}$: 1.788 Å Bond$_{34}$: 2.528 Å Bond$_{56}$: 2.514 Å Bond$_{78}$: 2.422 Å	-314.0 kJ/mol
IV	C_8				Bond$_{12}$: 2.357 Å Bond$_{34}$: 1.901 Å Bond$_{56}$: 2.298 Å	-381.7 kJ/mol
IV	C_9				Bond$_{12}$: 2.472 Å Bond$_{34}$: 1.869 Å Bond$_{56}$: 2.503 Å	-388.6 kJ/mol
IV	C_{10}				Bond$_{12}$: 2.382 Å Bond$_{34}$: 1.894 Å Bond$_{56}$: 2.299 Å Bond$_{78}$: 2.392 Å	-391.4 kJ/mol

Ca(CaO) O(CaO) As(As$_2$O$_3$) O(As$_2$O$_3$)

3.2. Adsorption Process

Due to the continuity of energy, the adsorption process can be characterized as an energy-drop process, including both physisorption and chemisorption.

3.2.1. Transformation Process of Physisorption Structures to Chemisorption Structures

In the following part, the transition state number n is abbreviated as TS$_n$, and the intermediate position number n is abbreviated as IP$_n$, for short.

As shown in Figure 1, when As$_2$O$_3$ approaches the surface with vibration along the surface, the physisorption structure transforms into a chemisorption structure during one or two transition state. For instance, P$_1$ to C$_7$ (Figure 1a), P$_2$ to C$_8$ (Figure 1b) and P$_3$ to C$_8$ (Figure 1c). The energy barrier is low, from 1.4 kJ/mol to 13.9 kJ/mol, suggesting that the physisorbed As$_2$O$_3$ is not stable enough and could be easily transformed into chemisorption structures by thermal vibration.

(a)

(b)

(c)

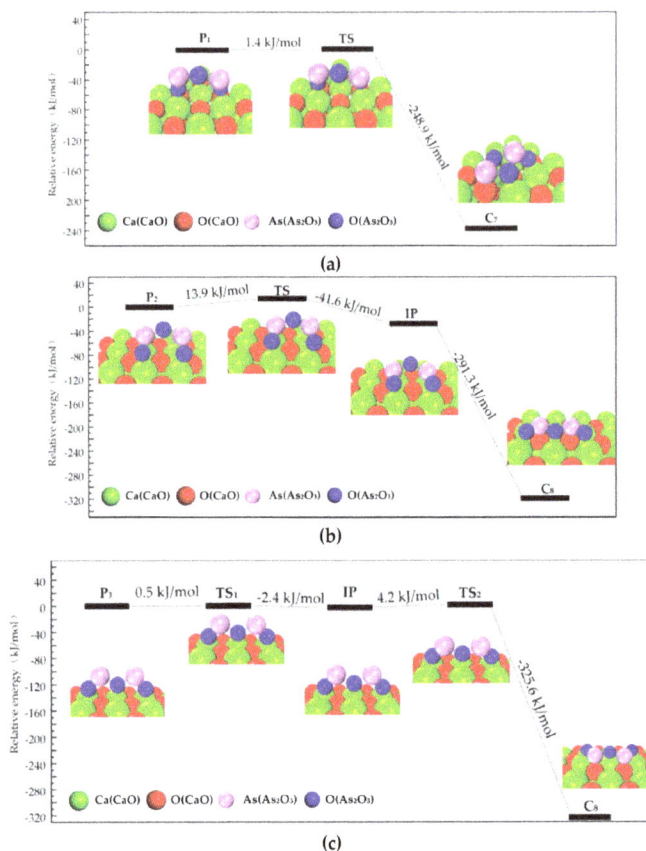

Figure 1. Structures and energies during transformation process of physisorption structures to chemisorption structures. (**a**) Reaction path of physisorption structure 1; (**b**) Reaction path of physisorption structure 2; (**c**) Reaction path of physisorption structure 3.

3.2.2. Transformation Process of Chemisorption Structures

Chemisorbed As_2O_3 gradually transforms into more stable structures. Different possible reaction paths were calculated. The four categories of chemisorption structures can be sorted by the E_{ads} of each as Category IV < Category III ≈ Category II < Category I.

Category I has relatively high energy, i.e., relatively low stability, its transformation to Category II, III and IV could be triggered by molecular thermal vibration.

The pathway that Category I transforms to Category II is shown in Figure 2. Firstly, C_1 transforms into C_6 (Category II) and then C_5 (Category III), with the energy barrier of 10.8 kJ/mol, 16.7 kJ/mol, and 6.7 kJ/mol, respectively. As shown in Figure 2, Category I transforms into Category IV along with another reaction path, the related energy barrier is 7.4 kJ/mol. The relatively low energy barrier suggests that Category I is not stable enough, and could easily transform to Category II, III and IV.

(a)

(b)

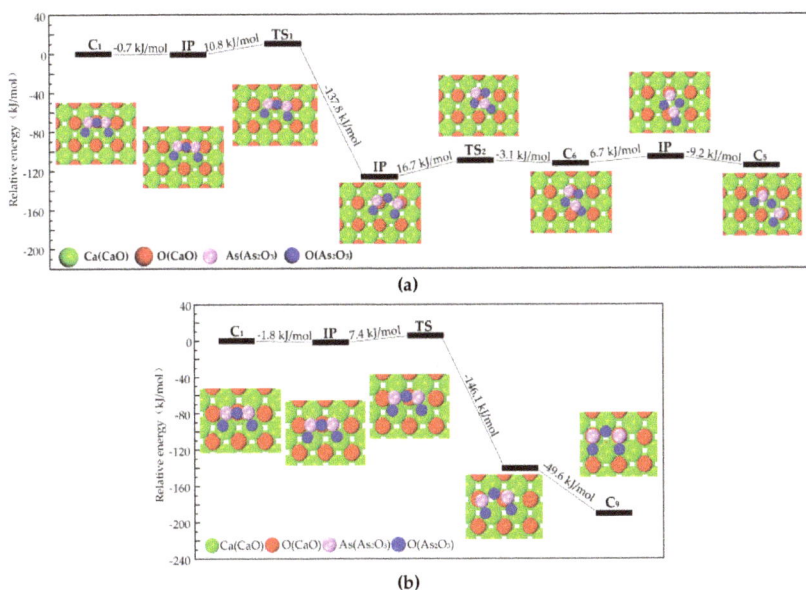

Figure 2. Transformation path of Category I. (**a**) Category I to Category II and III; (**b**) Category I to Category IV.

The reaction path of Category II is shown in Figure 3. C_3 firstly transforms into intermediate and then converts to C_9. The corresponding energy barrier is 16.1 kJ/mol and 83.0 kJ/mol, proving that Category II transforms to Category IV with the special direction of thermal vibration.

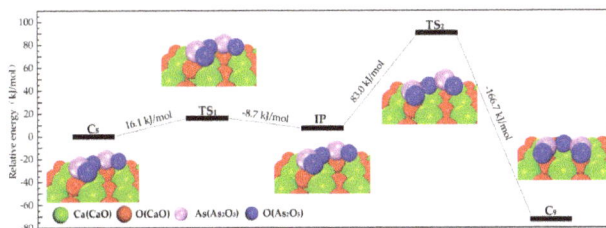

Figure 3. Transformation path of Category II.

Structures of Category III can transform into Category II, as shown in Figure 4. The spoon-shaped structure of As_2O_3 in C_7 disappears and then overcomes a 48.3 kJ/mol energy barrier to transform to C_5, indicating the conversion of Category III to Category II.

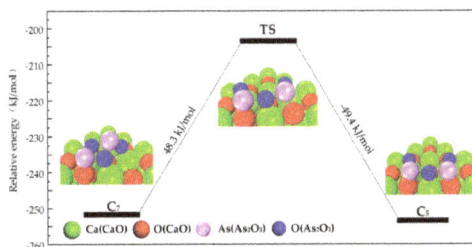

Figure 4. Transformation path of Category III.

Category IV is the most stable category. C_8, C_9, and C_{10} can transform into each other (shown in Figure 5). As_2O_3's As does not move during the transformation. When all of As_2O_3's O in C_8 vibrate, C_8 converts to C_9, and the energy barrier is 41.6 kJ/mol (Figure 5a). When one of As_2O_3's O in C_8 vibrates, C_8 converts to C_{10}, and the energy barrier is 153.3 kJ/mol (Figure 5b). When one of the oxygen atoms of As_2O_3 in C_9 vibrates, C_9 converts to C_{10}, and the energy barrier is 154.4 kJ/mol (Figure 5c). The difference in energy barrier is caused by the movement distance of As_2O_3's O being motivated by thermal vibration. In the first reaction, the movement distance of As_2O_3's O is shorter than that in the second and third reactions, which demands relatively lower energy to overcome the energy barrier.

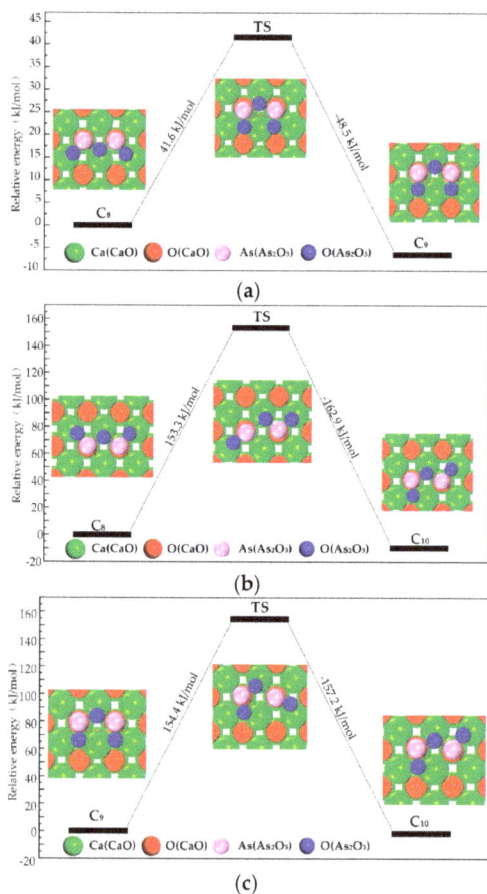

Figure 5. Transformation path of Category IV. (**a**) Reaction path of C_8 to C_9; (**b**) Reaction path of C_8 to C_{10}; (**c**) Reaction path of C_9 to C_{10}.

3.3. Path of the Reaction

According to above-mentioned processes, the reaction paths can be concluded as follows; firstly, the isolated As_2O_3 is physisorbed on a CaO surface (As_2O_3's O weakly interacts with superficial Ca); secondly, the physisorbed As_2O_3 transforms to chemisorbed As_2O_3. (As_2O_3's As interacts with superficial O); and thirdly, due to thermal vibration, the chemisorbed As_2O_3 transforms into more stable chemisorbed As_2O_3 (the position of As_2O_3's O changed).

The adsorption path of As_2O_3 was summarized as the process shown in Figure 6. These reactions could be classified as three types according to the energy barrier with the aim to reflect the intensity of the required reaction temperature. The number of superficial CaO occupied by As_2O_3 is also considered in order to describe the adsorption reaction equation.

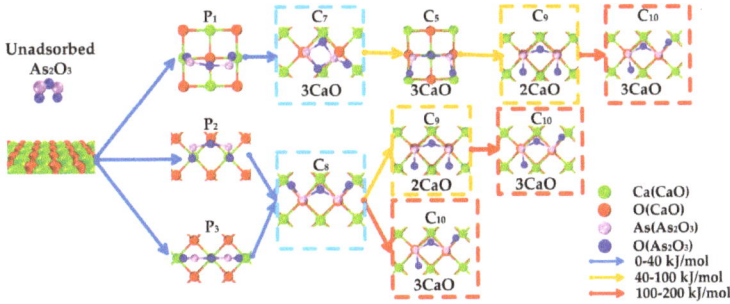

Figure 6. Overall adsorption paths of As_2O_3 on CaO.

Blue arrow: energy barrier is in the range of 0–40 kJ/mol, suggesting that reaction is likely to occur under a relatively low-temperature condition.

Yellow arrow: energy barrier is in the range of 40–100 kJ/mol, suggesting that reaction is likely to occur under a relatively medium-temperature condition.

Red arrow: energy barrier is in the range of 100–200 kJ/mol, suggesting that reaction is likely to occur under a relatively high-temperature condition.

Figure 6 reveals that three main reaction paths may exist:

1. $As_2O_3 \rightarrow P_1 \rightarrow C_7 \rightarrow C_9 \rightarrow C_{10}$;
2. $As_2O_3 \rightarrow P_2$ or $P_3 \rightarrow C_8 \rightarrow C_9 \rightarrow C_{10}$;
3. $As_2O_3 \rightarrow P_2$ or $P_3 \rightarrow C_8 \rightarrow C_{10}$.

Under a relatively low-temperature condition (blue arrow, 0–40 kJ/mol), the main products are C_7 and C_8 (blue grid). Three superficial Ca and one or two superficial O are involved in the reaction, representing three CaO participates in the adsorption. The adsorption equation could be written as:

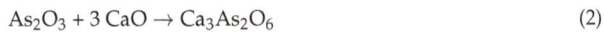

$$As_2O_3 + 3\,CaO \rightarrow Ca_3As_2O_6 \qquad (2)$$

Under a relatively medium-temperature condition (yellow arrow, 40–100 kJ/mol), the main products are C_9. Two superficial Ca and two superficial O participate in the structure. The adsorption equation could be written as:

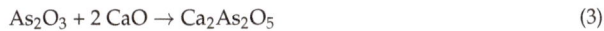

$$As_2O_3 + 2\,CaO \rightarrow Ca_2As_2O_5 \qquad (3)$$

Under a relatively high-temperature condition (red arrow, 100–200 kJ/mol), the main product is C_{10}. Three superficial Ca and two superficial O are involved in the reaction (hollow Ca represents 1/2 Ca atom). The adsorption equation could be written as:

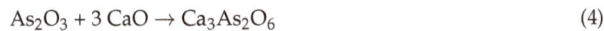

$$As_2O_3 + 3\,CaO \rightarrow Ca_3As_2O_6 \qquad (4)$$

With the reaction temperature increases, adsorption product changes from $Ca_3As_2O_6$ to $Ca_2As_2O_5$ and back to $Ca_3As_2O_6$ again. Different microcosmic adsorption structures lead to different macroscopic products and reaction equation.

Besides, as shown in Figure S2, the paths of C_1 transforming to other structures have been also been found. However, no possible paths which isolated or physisorbed As_2O_3 transforms to C_1 has been found, implying C_1 is unstable or nonexistent.

3.4. Partial Density of States (PDOS)

The change PDOS of As_2O_3 and CaO was put in the Supplementary Materials (Figure S3). For As_2O_3, the PDOS of physisorption structure 1, 2, and 3 are similar to each other. As the physisorption structure transforms to C_7, the p state orbitals near Fermi level (from -0.6 eV to 1.9 eV) drift to lower energy level, meanwhile get energy splitting and orbital reorganization, caused by the changing of As_2O_3 structure and the combination between As_2O_3's As and superficial O. When C_7 transforms to C_5, s state orbital (-17.2 eV) energy level splits into two peaks of -18.0 eV and -16.9 eV, which is caused by the As-O bond breaking and the bonding between As and superficial O. When C_5 transform to C_8, the p state orbital (3.7 eV) and s state orbital (-17.9 eV) energy level both split slightly. This is the result of the slight change in the surface distribution of As_2O_3. As the adsorption products have close energies and structures, PDOS of C_9 and C_{10} are basically similar to C_8.

For CaO slab surface, when an As_2O_3 molecule is physisorbed on the surface, little change of PDOS is detected. When As_2O_3 is chemisorbed, it can be seen that the superficial p orbitals around Fermi level (from -2.7 eV to 0.4 eV) drift to a lower energy range (from -5.8 eV to 0.2 eV). Moreover, a small peak (-16.8 eV) is separated from s orbitals (peak at -14.6 eV), proving that s orbitals participate in the chemisorption to some extent. Superficial p state orbitals near Fermi level play an important role in the chemisorption of As_2O_3. It suggests that the CaO surface's property of capturing As_2O_3 might be improved by increasing the quantities of superficial p orbitals near Fermi level.

3.5. Influence of O_2 on Adsorbed As_2O_3

Under the flue gas atmosphere, especially O_2-containing atmosphere, O_2 reacts with chemisorbed As_2O_3; i.e., arsenite (AsO_3^{3-}) is oxidized to arsenate (AsO_4^{3-}). As an example, two stable chemisorption structures (C_5, C_9) identified previously were presented in Table 4. The distance between As_2O_3's As and O_2's O is 1.763–1.764 Å, which is close to the As-O bond length of As_2O_3 (1.628 Å). The distance between O_2's O and superficial Ca is 2.247–2.263 Å. According to the electron density cloud, one of O_2's O overlaps with As_2O_3's As. The other O of O_2 overlaps slightly with the superficial Ca.

Table 4. Stable chemisorption structures under O_2 atmosphere, adsorption energy, electron density cloud and E_{ads}.

Name	Top View	Front View	Electron Density Cloud	Structure Details	E_{ads}
C_5 under O_2				$Bond_{12}$: 1.764 Å $Bond_{34}$: 1.452 Å $Bond_{56}$: 2.263 Å	-165.2 kJ/mol
C_8 under O_2				$Bond_{12}$: 1.763 Å $Bond_{34}$: 1.599 Å $Bond_{56}$: 2.427 Å	-174.4 kJ/mol

● Ca(CaO) ● O(CaO) ● As(As₂O₃) ● O(As₂O₃/O₂)

Based on Figure 6 and Table 4, the reaction equation of adsorption under O_2 atmosphere can be written as Equations (5)–(7), corresponding to low-temperature, medium-temperature, and high-temperature adsorption, respectively.

$$3CaO + As_2O_3 + O_2 \rightarrow Ca_3As_2O_8 \tag{5}$$

$$2CaO + As_2O_3 + O_2 \rightarrow Ca_2As_2O_7 \qquad (6)$$

$$3CaO + As_2O_3 + O_2 \rightarrow Ca_3As_2O_8 \qquad (7)$$

With the increase of reaction temperature, adsorption product changed from $Ca_3As_2O_8$ to $Ca_2As_2O_7$ and then to $Ca_3As_2O_8$ in an O_2-containing atmosphere. According to this research, the product under low-temperature and high-temperature conditions is $Ca_3As_2O_8$ with different structures, i.e., crystalline form. Under a medium-temperature condition, the main product is $Ca_3As_2O_7$.

Previous experimental research consistently reflected that the adsorption product with O_2 existence is AsO_4^{3-}, while different opinions existed regarding the adsorption structures. The study of Jadhav [12] found that the adsorption product obtained under 500 °C was mainly $Ca_3As_2O_8$ (JCPDS No.01-0933). Under 700 °C and 900 °C, the product was $Ca_2As_2O_7$ (JCPDS No.17-0444). When the temperature increased to 1000 °C, the reaction product was $Ca_3As_2O_8$ (JCPDS No.26-0295). Mahuli [41] (600 °C and 1000 °C) and Sterling [11] (800 °C) found that the adsorption product was $Ca_3As_2O_8$ (JCPDS No. 26-0295), while the sorbent used by Mahuli was $Ca(OH)_2$. Li [13] found that the product obtained under 600 °C mainly belonged to $Ca_3As_2O_8$ crystal structure (JCPDS No. 01-0933), and another kind of $Ca_3As_2O_8$ crystal (JCPDS No. 73-1928) was identified for the products obtained under 800 °C and 1000 °C.

The role of temperature on adsorption product transformation is qualitatively described. The more detailed description of the product layer development is associated with many other factors, such as the concentration and flow rate of As_2O_3 and O_2, and the quantity and granular size of CaO. The quantitative description of the adsorption process is still a very difficult challenge. Nevertheless, the DFT calculation findings revealed by this study could directly explain the experimental results obtained by previous researchers, which might provide some meaningful insight to understand the process of As_2O_3 adsorption on CaO.

4. Conclusions

The mechanisms of As_2O_3 adsorption on a CaO surface have been studied by using DFT calculation; conclusions are as follows:

(1) Physisorption active sites are composed of superficial Ca atoms that interact with O of As_2O_3. Chemisorption active sites are superficial O atoms that interact with As of As_2O_3;

(2) The adsorption process can be described as follows: the isolated As_2O_3 molecule is firstly adsorbed on the CaO surface by physisorption, and then physisorbed As_2O_3 will transform to chemisorbed As_2O_3. Due to thermal vibration, the chemisorbed As_2O_3 would overcome the energy barrier and transform to a more stable chemisorbed As_2O_3 state. The adsorption product is AsO_3^{3-};

(3) The adsorption products of As_2O_3 under an O_2-containing atmosphere are AsO_4^{3-}. The adsorption product's structure is influenced by the adsorption temperature. Under relatively low-temperature, the product is $Ca_3As_2O_8$; under relatively medium-temperature, the product is $Ca_3As_2O_7$; and under relatively high-temperature, the product is $Ca_3As_2O_8$.

The consistency between DFT calculation and the previous experiments proves high possibilities to design and optimized the CaO-based adsorbents by modifying O sites or other elements. Besides, other flue gases such as SO_2 or CO_2 can be involved in the following study to achieve materials design under real flue gas conditions. The optimized CaO-based adsorbents should be of high industrial value, could be applied in the injection of limestone into the furnace, CaO looping reactor, and dry desulfurization, etc.

Supplementary Materials: The following are available online at http://www.mdpi.com/1996-1944/12/4/677/s1. Table S1: Changes in physical and chemical properties of different surface size; Table S2: Changes in physical and chemical properties of different layers; Figure S1: Initial adsorbate structures; Figure S2: Paths and structures of the physisorption and chemisorption reaction from chemisorption structure 1; Figure S3: PDOS of As_2O_3 and CaO surface during physisorption and chemisorption (a. PDOS of As_2O_3 molecule; and b. PDOS of CaO surface).

Author Contributions: Research Design, Y.F. and Y.Z.; Data collection, Y.F.; Data analysis, Y.F., Q.W., S.D., P.H. and D.L.; Figures and Tables, Q.W.; Manuscript draft, Y.F., Q.W., Y.Z. and S.D.; Manuscript revise, P.H. and D.L.

Funding: This work was financially supported by the National Natural Science Foundation of China No. 51776107.

Conflicts of Interest: The authors declare no conflict of interest.

References

1. Liu, R.; Yang, J.; Xiao, Y.; Liu, Z. Fate of Forms of arsenic in Yima coal during pyrolysis. *Energy Fuels* **2009**, *23*, 2013–2017. [CrossRef]
2. Clarke, L.B.; Sloss, L.L. *Trace Elements-Emissions from Coal Combustion and Gasification*; IEA Coal Research: London, UK, 1992; Volume 8, pp. 1822–1826.
3. Gao, J.; Yu, J.; Yang, L. Urinary arsenic metabolites of subjects exposed to elevated arsenic present in coal in Shaanxi province, China. *Int. J. Environ. Res. Public Health* **2011**, *8*, 1991–2008. [CrossRef] [PubMed]
4. Kapaj, S.; Peterson, H.; Liber, K.; Bhattacharya, P. Human health effects from chronic arsenic poisoning—A review. *J. Environ. Sci. Health Part A* **2006**, *41*, 2399–2428. [CrossRef] [PubMed]
5. 1990 Clean Air Act Amendment. Public Law; 1990. Available online: https://www.gpo.gov/fdsys/pkg/USCODE-2013-title42/html/USCODE-2013-title42-chap85-subchapI-partA-sec7412.htm (accessed on 11 February 2019).
6. Duan, J.; Tan, J.; Hao, J.; Chai, F. Size distribution, characteristics and sources of heavy metals in haze episod in Beijing. *J. Environ. Sci.* **2014**, *26*, 189–196. [CrossRef]
7. Pacyna, J.M.; Pacyna, E.G. An assessment of global and regional emissions of trace metals to the atmosphere from anthropogenic sources worldwide. *Environ. Rev.* **2001**, *9*, 269–298. [CrossRef]
8. Tian, H.; Liu, K.; Zhou, J.; Lu, L.; Hao, J.; Qiu, P.; Gao, J.; Zhu, C.; Wang, K.; Hua, S. Atmospheric Emission Inventory of Hazardous Trace Elements from China's Coal-Fired Power Plants Temporal Trends and Spatial Variation Characteristics. *Environ. Sci. Technol.* **2014**, *48*, 3575–3582. [CrossRef] [PubMed]
9. Mercury and air toxics Standards (MATS). Public Law; 2013. Available online: https://www.epa.gov/mats (accessed on 11 February 2019).
10. Helsen, L. Sampling technologies and air pollution control devices for gaseous and particulate arsenic: A review. *Environ. Pollut.* **2005**, *137*, 305–315. [CrossRef] [PubMed]
11. Sterling, R.O.; Helble, J.J. Reaction of arsenic vapor species with fly ash compounds: Kinetics and speciation of the reaction with calcium silicates. *Chemosphere* **2003**, *51*, 1111–1119. [CrossRef]
12. Jadhav, R.A.; Fan, L. Capture of gas-phase arsenic oxide by lime: Kinetic and mechanistic studies. *Environ. Sci. Technol.* **2001**, *35*, 794–799. [CrossRef] [PubMed]
13. Li, Y.; Tong, H.; Zhuo, Y.; Li, Y.; Xu, X. Simultaneous removal of SO_2 and trace As_2O_3 from flue gas: Mechanism, kinetics study, and effect of main gases on arsenic capture. *Environ. Sci. Technol.* **2007**, *41*, 2894–2900. [CrossRef] [PubMed]
14. Li, Y. *Experimental Study on Simultaneous Removal of Trace Selenium and Arsenic in Flue Gas Desulphurization within Medium Temperature Range*; Tsinghua University: Beijing, China, 2006.
15. López-Antón, M.A.; Díaz-Somoano, M.; Spears, D.A.; Martínez-Tarazona, M.R. Arsenic and selenium capture by fly ashes at low temperature. *Environ. Sci. Technol.* **2006**, *40*, 3947–3951. [CrossRef] [PubMed]
16. Shah, P.; Strezov, V.; Stevanov, C.; Nelson, P.F. Speciation of arsenic and selenium in coal combustion products. *Energy Fuels* **2007**, *21*, 506–512. [CrossRef]
17. Huggins, F.E.; Senior, C.L.; Chu, P.; Ladwig, K.; Huffman, G.P. Selenium and arsenic speciation in fly ash from full-scale coal-burning utility plants. *Environ. Sci. Technol.* **2007**, *41*, 3284–3289. [CrossRef] [PubMed]
18. Tian, C.; Gupta, R.; Zhao, Y.; Zhang, J. Release Behaviors of Arsenic in Fine Particles Generated from a Typical High-Arsenic Coal at a High Temperature. *Energy Fuels* **2016**, *30*, 6201–6209. [CrossRef]
19. Jialin, Y.; Jingjing, X.; Qinfang, Z.; Binwen, Z.; Baolin, W. First-principles studies on the structural and electronic properties of As clusters. *Materials* **2018**, *11*, 1596.
20. Li, Z.; Yangwen, W.; Jian, H.; Qiang, L.; Yongping, Y.; Laibao, Z. Mechanism of mercury adsorption and oxidation by oxygen over the CeO_2 (111) surface: A DFT study. *Materials* **2018**, *11*, 485.

21. Zhang, S.; Hu, X.; Lu, Q.; Zhang, J. Density functional theory study of arsenic and selenium adsorption on the CaO (001) surface. *Energy Fuels* **2011**, *25*, 2932–2938. [CrossRef]

22. Segall, M.D.; Lindan, P.J.; Probert, M.A.; Pickard, C.J.; Hasnip, P.J.; Clark, S.J.; Payne, M.C. First-principles simulation: Ideas, illustrations and the CASTEP code. *J. Phys. Condens. Matter* **2002**, *14*, 2717. [CrossRef]

23. Clark, S.J.; Segall, M.D.; Pickard, C.J.; Hasnip, P.J.; Probert, M.I.; Refson, K.; Payne, M.C. First principles methods using CASTEP. *Cryst. Mater.* **2005**, *220*, 567–570. [CrossRef]

24. Burke, K.; Ernzerhof, M.; Perdew, J.P. Generalized Gradient Approximation Made Simple. *Phys. Rev. Lett.* **1996**, *77*, 3865–3868.

25. Vanderbilt, D. Soft self-consistent pseudopotentials in a generalized eigenvalue formalism. *Phys. Rev. B Condens. Matter* **1990**, *41*, 7892. [CrossRef] [PubMed]

26. Von Barth, U.; Hedin, L. A local exchange-correlation potential for the spin polarized case. i. *J. Phys. C Solid State Phys.* **1972**, *5*, 1629. [CrossRef]

27. Press, W.H.; Teukolsky, S.A.; Vetterling, W.T.; Flannery, B.P. *Numerical Recipes in C.*; Cambridge University Press: Cambridge, UK, 1996; Volume 2.

28. Halgren, T.A.; Lipscomb, W.N. The synchronous-transit method for determining reaction pathways and locating molecular transition states. *Chem. Phys. Lett.* **1977**, *49*, 225–232. [CrossRef]

29. Henkelman, G.; Jónsson, H. Improved tangent estimate in the nudged elastic band method for finding minimum energy paths and saddle points. *J. Chem. Phys.* **2000**, *113*, 9978–9985. [CrossRef]

30. Monkhorst, H.J.; Pack, J.D. Special points for Brillouin-zone integrations. *Phys. Rev. B Condens. Matter* **1976**, *13*, 5188. [CrossRef]

31. Ghebouli, M.A.; Ghebouli, B.; Bouhemadou, A.; Fatmi, M.; Bouamama, K. Structural, electronic, optical and thermodynamic properties of SrxCa1−xO, BaxSr1−xO and BaxCa1−xO alloys. *J. Alloys Compd.* **2011**, *509*, 1440–1447. [CrossRef]

32. Verbraeken, M.C.; Suard, E.; Irvine, J.T. Order and disorder in Ca2ND0.90H0.10–A structural and thermal study. *J. Solid State Chem.* **2011**, *184*, 2088–2096. [CrossRef]

33. Da Hora, G.C.A.; Longo, R.L.; Da Silva, J.B.P. Calculations of structures and reaction energy profiles of As_2O_3 and As_4O_6 species by quantum chemical methods. *Int. J. Quantum Chem.* **2012**, *112*, 3320–3324. [CrossRef]

34. Liu, H.; Xiang, H.; Gong, X.G. First principles study of adsorption of O_2 on Al surface with hybrid functionals. *J. Chem. Phys.* **2011**, *135*, 214702. [CrossRef] [PubMed]

35. Fan, Y.; Yao, J.G.; Zhang, Z.; Sceats, M.; Zhuo, Y.; Li, L.; Maitland, G.C.; Fennell, P.S. Pressurized calcium looping in the presence of steam in a spout-fluidized-bed reactor with dft analysis. *Fuel Process. Technol.* **2018**, *169*, 24–41. [CrossRef]

36. Fan, Y.; Zhuo, Y.; Zhu, Z.; Du, W.; Li, L. Zerovalent selenium adsorption mechanisms on a CaO surface: DFT calculation and experimental study. *J. Phys. Chem. A* **2017**, *121*, 7385. [CrossRef] [PubMed]

37. Fan, Y.; Zhuo, Y.; Li, L. SeO_2 adsorption on a CaO surface: DFT and experimental study on the adsorption of multiple SeO_2 molecules. *Appl. Surf. Sci.* **2017**, *420*, 465–471. [CrossRef]

38. Sun, S.; Zhang, D.; Li, C.; Wang, Y.; Yang, Q. Density functional theory study of mercury adsorption and oxidation on CuO (111) surface. *Chem. Eng. J.* **2014**, *258*, 128–135. [CrossRef]

39. Xiang, W.; Liu, J.; Chang, M.; Zheng, C. The adsorption mechanism of elemental mercury on CuO (110) surface. *Chem. Eng. J.* **2012**, *200*, 91–96. [CrossRef]

40. Nørskov, J.K.; Studt, F.; Abild-Pedersen, F.; Bligaard, T. *Fundamental Concepts in Heterogeneous Catalysis*; John Wiley & Sons: Hoboken, NJ, USA, 2014.

41. Mahuli, S.; Agnihotri, R.; Chauk, S.; Ghosh-Dastidar, A.; Fan, L.-S. Mechanism of Arsenic Sorption by Hydrated Lime. *Environ. Sci. Technol* **1998**, *31*, 3226–3231. [CrossRef]

materials

MDPI

Article

A Multiscale Model of Oxidation Kinetics for Cu-Based Oxygen Carrier in Chemical Looping with Oxygen Uncoupling

Hui Wang [1,2], Zhenshan Li [1,2,*] and Ningsheng Cai [1,2]

[1]   Key Laboratory for Thermal Science and Power Engineering of Ministry of Education, Department of Energy and Power Engineering, Tsinghua University, Beijing 100084, China; thu_wh@126.com (H.W.); cains@tsinghua.edu.cn (N.C.)
[2]   Tsinghua University-University of Waterloo Joint Research Center for Micro/Nano Energy & Environment Technology, Tsinghua University, Beijing 100084, China
*   Correspondence: lizs@tsinghua.edu.cn

Received: 23 January 2019; Accepted: 8 April 2019; Published: 10 April 2019

Abstract: Copper oxide is one of the promising oxygen carrier materials in chemical looping with oxygen uncoupling (CLOU) technology, cycling between Cu_2O and CuO. In this study, a multiscale model was developed to describe the oxidation kinetics of the Cu-based oxygen carrier particle with oxygen, including surface, grain, and particle scale. It was considered that the solid product grows with the morphology of disperse islands on the grain surface, and O_2 contacts with two different kinds of grain surfaces in the grain scale model, that is, Cu_2O surface (solid reactant surface) and CuO surface (solid product surface). The two-stage behavior of the oxidation reaction of the Cu-based oxygen carrier was predicted successfully using the developed model, and the model results showed good agreement with experimental data in the literature. The effects of oxygen partial pressure, temperature, and particle structure on the oxidation performance were analyzed. The modeling results indicated that the transition of the conversion curve occurs when product islands cover most part of the grain surface. The oxygen partial pressure and particle structure have an obvious influence on the duration time of the fast reaction stage. Furthermore, the influence of the external mass transfer and the change of effectiveness factor during the oxidation reaction process were discussed to investigate the controlling step of the reaction. It was concluded that the external mass transfer step hardly affects the reaction performance under the particle sizes normally used in CLOU. The value of the effectiveness factor increases as the reaction goes by, which means the chemical reaction resistance at grain scale increases resulting from the growing number of product islands on the grain surface.

Keywords: oxygen carrier; multiscale model; product island; oxidation kinetics

1. Introduction

Chemical looping combustion (CLC) is a new combustion technology [1,2], where oxygen carriers are used to transport oxygen from the air reactor to the fuel reactor through the redox cycle. Compared with traditional CO_2 capture technologies, such as pre-combustion capture [3], post-combustion capture [4], and oxy-fuel combustion [5], CLC technology has obvious advantages in reducing the energy consumption of CO_2 capture. The chemical looping with oxygen uncoupling (CLOU) concept [6] is based on CLC technology, where the oxygen carriers have oxygen release capacity. Solid fuel can react directly with oxygen released from oxygen carriers in the fuel reactor to improve combustion efficiency. The research results of Mattisson et al. [6] show that when petroleum coke is used as fuel, the conversion of the CLOU process is 50 times higher than that of traditional CLC process. Subsequently, many researchers further explored the oxygen carriers suitable for CLOU technology [7–16].

The Cu-based oxygen carrier was reported to have a strong oxygen release capacity and fast reaction rate in other research [9–16], and the corresponding redox pair was CuO/Cu$_2$O. Both the oxidation and reduction of Cu-based oxygen carriers are gas-solid reactions. There are a large number of studies on the oxidation or reduction kinetics of Cu-based oxygen carriers. de Diego [11], Goldstein [12], and Gayn [13] used pure CuO as oxygen carriers to conduct the kinetic tests. In addition, researchers [14–16] used inert carrier materials and preparation methods to prepare Cu-based oxygen carriers with improved cyclic stability. It was widely found in the experimental results that there is a transition of the kinetics from the initial fast stage to the second slower stage in the conversion curve of the Cu$_2$O oxidation reaction [11,13–16].

To explain the kinetic behavior of the gas-solid reaction kinetics of the oxygen carrier, many kinds of models were developed. In the research of Clayton et al. [16], two apparent models, pore-blocking model and Avrami-Erofeev model, were used for the oxidation reaction of Cu$_2$O in the lower temperature range (below 700 °C) and higher temperature range (above 800 °C), respectively. García-Labiano et al. [17] and Maya et al. [18] used the grain model to predict the reaction behavior of oxygen carriers. The grain model assumes the particle to be a spherical porous solid particle that consists of numerous small grains within, and each of these grains is described using the unreacted shrinking core model. In addition, Dennis et al. [19] and Liu et al. [20] utilized the pore model to explain the kinetic performance of gas-solid reaction. Nevertheless, the models mentioned above were all focused on the particle scale step, including the external and internal mass transfer, while these models did not consider the microscopic reaction steps taking place on the grain surface and the growth of the solid product, which play an important role in the gas-solid reaction process [21]. Therefore, these models cannot explain the transition phenomenon from the initial fast stage to the second slower stage in the conversion curve from the view of the microscopic reaction process.

In the review paper of Gattinoni et al. [22], recent surface science, spectroscopy, and atomic computation work performed to understand the copper oxidation from the microscopic point of view was summarized and discussed. A good amount of computational work has been performed on the formation of copper oxides, providing important information on surface reaction process. However, few experimental studies are available to either confirm or disprove some computational results obtained at the atomic scale, and the nucleation details of the oxide islands are still unknown. Also, Zhang et al. [23] and Yu et al. [24] applied density functional theory to investigate the oxygen adsorption and dissociation process on the Cu$_2$O surfaces. The calculated results showed that the presence of oxygen vacancy on the surface exhibited a strong chemical reactivity towards the dissociation of O$_2$. Recently, it was found that the solid product showed dispersed and three-dimensional morphology on the solid reactant surface [25–28]. In addition, a rate equation theory for Fe oxidation was developed to describe the nucleation and growth process of the solid product [25]. However, the above research at microscopic scale did not consider the steps involved at the particle scale, such as particle structure change, external mass transfer, and internal mass transfer, thus could not explain the phenomenon at the macroscopic scale and the controlling mechanism of the reaction.

The gas-solid reaction of the Cu-based oxygen carrier with oxygen is a multiscale behavior. It is necessary to study the reaction process from multiscale points of view. In this study, a multiscale model, including surface, grain, and particle scale, was established to describe the oxidation reaction of Cu-based oxygen carrier, and the developed model was validated with experimental data in the literature. Then, the model was used to analyze the effects of oxygen partial pressure, temperature, and particle structure on the oxidation reaction behaviors. Further, the controlling step of the oxidation reaction of the Cu-based oxygen carrier particle was discussed.

2. Mathematical Model

In this study, a multiscale model was developed to describe the oxidation reaction of the Cu-based oxygen carrier. The oxidation reaction of the Cu-based oxygen carrier is

$$Cu_2O + \frac{1}{2}O_2 \rightarrow 2CuO \tag{1}$$

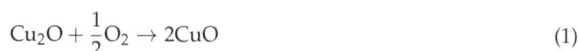

As shown in Figure 1, the oxidation process of the oxygen carrier particle was described at multiscale, that is, surface scale, grain scale, and particle scale. The oxygen carrier particle was considered as spherical porous media, which is composed of a matrix of spherical nonporous grains, and O_2 could diffuse into the particle. The interaction of O_2 with Cu_2O occurs on the grain surface, and the solid product will grow on the grain surface, which can cover the grain surface and result in the change of grain size and an increase of gas diffusion resistance. In the thermogravimetric analysis (TGA) experiments of Clayton et al. [16], it was pointed out that the Cu-based oxygen carrier particle could be considered isothermal during the oxidation process. In addition, in the study of García-Labiano et al. [17], the coupled energy equations were considered in their grain model, and it was also concluded that the oxygen carrier particle could be considered isothermal for most of the reactions in a chemical looping combustion process. Therefore, based on the above studies, the temperature inside the oxygen carrier particle is considered isothermal in the model of this work.

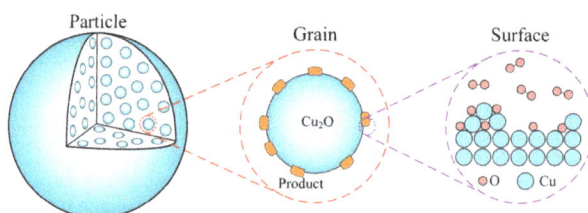

Figure 1. Schematic diagram of the multiscale model.

2.1. Model at Surface Scale

Up to now, many theoretical and experimental studies have been reported to investigate the O_2 adsorption on Cu-based metal oxide surface [22–24,29–34]. When the Cu_2O surface is exposed to oxygen, oxygen molecules would be absorbed on the surface and generate adsorbed oxygen and oxygen ions [29–34]. On the Cu_2O-O_2 interface, the surface reaction involves several steps [31–33]: (R1) O_2 gas molecules are adsorbed on the metal oxide surface and decomposed into adsorbed oxygen, O(ads); (R2) O(ads) takes electrons from the metal cations (Cu^+) and forms chemically adsorbed oxygen, O^-(chem); (R3) O^-(chem) further reacts with electrons to form lattice oxygen, O^{2-}(latt). During the reaction, the oxygen from the most outward lattice position would diffuse in the crystal structure to replenish oxygen vacancies. This reaction process can be described as follows.

$$O_2(g) + 2* \underset{k_{-1}}{\overset{k_1}{\rightleftharpoons}} 2O(ads) \tag{R1}$$

$$2O(ads) + 2e^- \underset{k_{-2}}{\overset{k_2}{\rightleftharpoons}} 2O^-(chem) \tag{R2}$$

$$2O^-(chem) + 2e^- \overset{k_3}{\rightarrow} 2O^{2-}(latt) \tag{R3}$$

The reaction rate of step (R1)–(R3) can be given by

$$r_{R1} = k_1\left(P_{O_2} - P_e\right)\theta_s^2 - k_{-1}\theta_{O(ads)}^2 \tag{2}$$

$$r_{R2} = k_2\theta_{O(ads)}^2 - k_{-2}\theta_{O(chem)}^2 \tag{3}$$

$$r_{R3} = k_3\theta_{O(chem)}^2 \tag{4}$$

where k_i, $i = \pm 1, \pm 2, 3$ is the reaction rate constant, and the surface site densities are lumped in the reaction rate constant. $\theta_s, \theta_{O(ads)}, \theta_{O(chem)}$ are the coverage ratios of active sites, O(ads), O^-(chem) on the surface, respectively, which satisfy

$$\theta_s + \theta_{O(ads)} + \theta_{O(chem)} = 1 \tag{5}$$

P_e is the equilibrium partial pressure of O_2, which is expressed as [16]

$$P_e = 6.057 \cdot 10^{-11} \exp[0.02146(T - 273)] \tag{6}$$

Assuming that R1 and R2 are in chemical equilibrium, r_{R1} and r_{R2} are much less than r_{R3}, and hardly zero. Therefore, the reaction rate of R3 could be obtained by combining Equations (2)–(5):

$$r_{R3} = \frac{k_3 \frac{k_1}{k_{-1}} \frac{k_2}{k_{-2}} (P_{O_2} - P_e)}{\left[1 + \sqrt{\frac{k_1}{k_{-1}} (P_{O_2} - P_e)} + \sqrt{\frac{k_1}{k_{-1}} \frac{k_2}{k_{-2}} (P_{O_2} - P_e)}\right]^2} \tag{7}$$

In the case of $\frac{k_1}{k_{-1}} (P_{O_2} - P_e) \ll 1$, there is

$$r_{R3} = k(C - C_e), \quad k = k_3 \frac{k_1}{k_{-1}} \frac{k_2}{k_{-2}} R_g T \tag{8}$$

where C is the oxygen gas concentration, C_e is the equilibrium oxygen concentration, R_g is the gas constant, and T is the temperature.

The reaction rate constants involved in the expression of r_{R3} can be calculated through the atomic computation, as discussed in the review paper of Gattinoni et al. [22]. The surface scale model described above could provide the link between microscopic surface reaction and macroscopic kinetics. However, the atomic computation on the reaction rate constants is not the focus of this work. The value of k is considered as an adjustable parameter in this work.

2.2. Model at Grain Scale

In the traditional grain models [17,18], the solid product is assumed to form and grow in a uniform layer-by-layer mode on the grain surface, and the morphology of the solid product is a nonporous film that covers the unreacted core. The theory of a critical product layer thickness [35] is now used in most grain models to explain the end of the fast reaction period. However, because of the simplified description of solid product formation and growth involved in the initial stage, the traditional grain models cannot explain the kinetics well. The typical conversion curve of oxidation of the Cu-based oxygen carrier shows the two-stage shape. The initial stage of the oxidation reaction is fast, and a high conversion of the particle will be achieved in a relatively short time. The fast stage is followed by a slower stage, where the conversion increases slowly. Because of the assumption of uniform solid product film growth, the traditional grain model cannot describe the transition behavior of the reaction kinetics, as shown in Figure 2a.

By using atomic force microscopy (AFM) and a single crystal sample, it has been proposed in several studies that during the gas-solid reaction process, the solid product grows as dispersed, three-dimensional islands rather than a uniform continuous product layer [25–28]. All islands show similar shape, which can be explained by the Wuff construction theory [36], and the size of islands increases with the oxidation reaction time [25]. Recently, a rate equation theory of metal oxidation was developed to calculate the solid product nucleation and growth rates [25]. It was proposed that the critical size of product islands is temperature-dependent [37]. Therefore, the assumption of a uniform product film on the grain surface was replaced with the product island morphology in the model, as shown in Figure 2a. As the reaction goes on, the product islands will cover the grain surface and hinder the direct contact of the solid reactant with O_2, as shown in Figure 2b, which slows down the reaction rate significantly and results in the transition of the kinetics from the fast stage to the second slower stage.

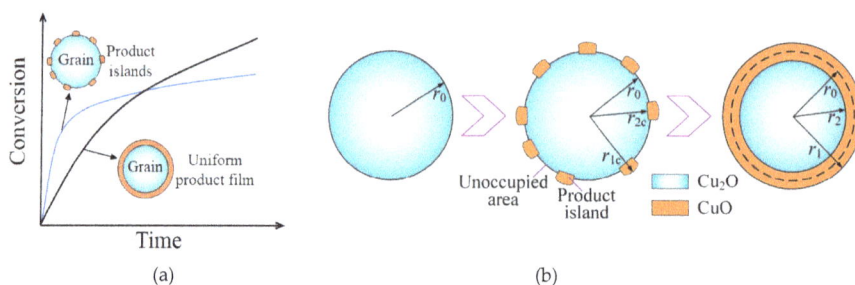

(a)   (b)

Figure 2. (a) Comparison of the product-islands-based model and traditional grain model; (b) Schematic diagram of the solid product morphology on grain surface during the oxidation reaction.

It is considered that O_2 contacts with two different kinds of grain surfaces in the model, that is, Cu_2O surface (solid reactant surface) and CuO surface (solid product surface) [22]. In the case of O_2 contacting Cu_2O surface, the reaction described in Section 2.1 will take place directly on this surface, and product islands will grow on the grain surface meanwhile. The size of the new product island was considered as a critical value of r_{2c} and r_{1c}, as shown in Figure 2b, where r_{2c} is the critical grain radius of Cu_2O/CuO interface and r_{1c} is the critical grain radius of CuO/O_2 interface. The critical solid reactant layer thickness is described as $h_c = r_0 - r_{2c}$, where r_0 is the initial grain radius. A key parameter at the grain scale in the model is the ratio of the unoccupied area on the grain surface, which is denoted as δ. By considering the reaction rate of solid reactant and the change rate of the unoccupied area, the expression of δ can be given as [38]

$$\delta = \exp\left[\frac{-2V_{Cu_2O}^M k(C - C_e)}{h_c} t\right] \qquad (9)$$

where $V_{Cu_2O}^M$ is the molar volume of Cu_2O (23.87 cm^3/mol), h_c is the critical solid reactant layer thickness, and t is the reaction time.

In the case of O_2 contacting solid product, the charged particles diffusion through the product layer will occur, which can be explained by the Wagner theory [39]. According to the Wagner theory, the diffusion flow rate of metal cations moving through the product layer should be equal to the diffusion flow rate of anions and electrons moving through the product layer to ensure charge conservation. The reaction of metal cations (Cu^+) and oxygen happens on the Cu_2O/CuO interface. As the reaction goes on, the size of the solid product will increase and be larger than the critical size, as shown in Figure 2b, where r_2 is the grain radius of Cu_2O/CuO interface and r_1 is the grain radius of CuO/O_2 interface. In addition, the number of product islands will increase and finally product islands will cover the grain surface completely. By integrating the diffusion flow rate of metal cations (Cu^+) through the product layer with the change rate of grain size, the change rate of grain size can be obtained as [38]

$$\frac{dr_2}{dt} = -\frac{2V_{Cu_2O}^M \frac{r_1^2}{r_2^2} k(C - C_e)}{r_1\left(\frac{r_1}{r_2} - 1\right)\frac{k(C-C_e)}{D_s C_{met}} + 1} \qquad (10)$$

where D_s is the diffusivity of metal ions (Cu^+) through the product layer, and C_{met} is the concentration of Cu^+ on the Cu_2O/CuO interface, which was considered to remain unchanged during the reaction. The initial condition for Equation (10) is $r_2 = r_{2c}$. The grain size, r_1, is calculated with the following equation:

$$r_1^3 = r_2^3 + \Pi\left(r_0^3 - r_2^3\right) \qquad (11)$$

where $Z = 2V_{CuO}^M/V_{Cu_2O}^M$ is the stoichiometric molar volume ratio of the solid product to the solid reactant, and V_{CuO}^M is the molar volume of CuO (12.52 cm^3/mol).

The local conversion at each position inside the particle is calculated as

$$\alpha = (1-\delta)\left(1 - \frac{r_2^3}{r_0^3}\right) \tag{12}$$

The local porosity inside the particle during the reaction is calculated as a function of the initial porosity and the local conversion:

$$\varepsilon = 1 - (1-\varepsilon_0) \cdot [1 + (Z-1) \cdot \alpha] \tag{13}$$

where ε is the local porosity, ε_0 is the initial porosity, and α is the local conversion.

2.3. Model at Particle Scale

Considering the external gas diffusion, internal gas diffusion, and chemical reaction, the O_2 concentration profile inside the particle can be obtained by making a mass balance for the particle:

$$\frac{1}{R^2}\frac{\partial}{\partial R}\left(R^2 D_e \frac{\partial C}{\partial R}\right) = \frac{(1-\varepsilon_0)}{2V_{Cu_2O}^M}\frac{\partial \alpha}{\partial t} \tag{14}$$

where R is the particle radius. The boundary conditions are as follows:

$$\left.\frac{\partial C}{\partial R}\right|_{R=0} = 0 \tag{15}$$

$$-D_e \left.\frac{\partial C}{\partial R}\right|_{R=R_0} = k_g(C_s - C_0) \tag{16}$$

where R_0 is the initial particle radius, and C_s is the concentration of O_2 on the particle external surface. The external mass transfer coefficient, k_g, is expressed as a function of Sherwood number:

$$Sh = \frac{2k_g R_0}{D_{O_2}} = 2 + 0.664 Re^{1/2} Sc^{1/3} \tag{17}$$

where Sh is the Sherwood number, Re is the Reynolds number, and Sc is the Schmidt number. D_{O_2} is the gas molecular diffusivity and is given using the equation developed by Fuller et al. [40]:

$$D_{O_2} = \frac{10^{-3}T^{1.75}\left(\sum_{i=1}^N \frac{1}{M_i}\right)^{0.5}}{P\left[\sum_{i=1}^N (v_i)^{1/3}\right]^2} \tag{18}$$

where M is the molar mass of the molecule, P is the pressure, and v is the diffusion volume for the molecule.

The effective gas diffusivity D_e inside the porous particle is expressed as

$$D_e = D_1 \cdot \varepsilon^2 \tag{19}$$

The gas diffusivity, D_1, is calculated as a combination of the gas molecular diffusivity and Knudsen diffusivity:

$$D_1 = \left(D_{O_2}^{-1} + D_K^{-1}\right)^{-1} \tag{20}$$

The Knudsen diffusivity, D_K, describes the collision of the gas molecules with the pore wall and is calculated as [41]

$$D_K = \frac{4}{3}\left(\frac{8R_g T}{\pi \cdot MW_{O_2}}\right)^{1/2}\left[\frac{128}{9}\cdot\frac{3(1-\varepsilon_0)}{4\pi r_0{}^3}r_0{}^2\left(1+\frac{\pi}{8}\right)\right]^{-1} \tag{21}$$

where R_g is the universal gas constant. The overall particle conversion, $\bar{\alpha}$, can be obtained by integrating all local conversions:

$$\bar{\alpha} = \frac{3}{R_0{}^3}\int_0^{R_0} R^2\alpha\cdot dR \tag{22}$$

3. Results

3.1. Effects of O_2 Partial Pressure

The equilibrium curve of oxygen partial pressure in the oxidation reaction can be obtained using Equation (6), as shown in Figure 3. In high-O_2 environments (such as the air reactor of a CLOU system), the Cu_2O tends to be oxidized by O_2, generating CuO. The driving force, which is defined as the difference between the actual partial pressure of oxygen and equilibrium partial pressure of oxygen [16], will affect the oxidation reaction rate of the oxygen carrier particle with O_2. When the oxygen particle is in an atmosphere of air, the driving force will decrease with an increase in the temperature.

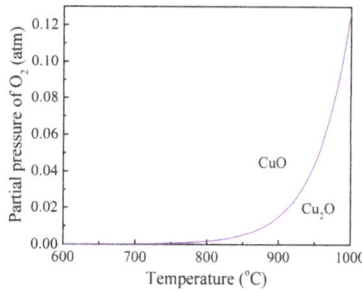

Figure 3. Equilibrium partial pressure of oxygen over the CuO/Cu_2O redox pair.

The experimental data of the TGA in the study of Adánez-Rubio et al. [42] were used to validate the developed model in this study and analyze the effects of O_2 partial pressure on the oxidation kinetics of Cu_2O. In their experiment, the particles, prepared through calcination of 24 h at 1100 °C, had an average particle size of 200 μm and a porosity of 0.161, and the oxidation experiments were conducted at oxygen concentration values from 2.5 to 21 vol.% at 900 °C. The experimental data and modeling results were compared, as shown in Figure 4. It can be seen that the modeling results agree well with the experimental data and the developed multiscale model can predict the transition of the reaction kinetics from the initial fast stage to the second slower stage successfully. It is clear that the supplied oxygen partial pressure affects the particle conversion significantly. The oxidation reaction rate is fast when the oxygen partial pressure is much higher than the equilibrium oxygen partial pressure but decreases quickly with the decrease in the driving force. Whitty et al. [43] also observed this phenomenon.

The profile of the ratio of the unoccupied area on the grain surface was plotted, as shown in Figure 5. It indicates that the transition of the conversion curve in Figure 4 happens when the product islands cover the grain surface. As observed in Figure 4, the duration time of the fast reaction stage will increase with the decrease in the supplied oxygen partial pressure. The reason is that the change rate of the ratio of the unoccupied area on the grain surface decreases with the decrease in the supplied oxygen partial pressure, as described in Equation (9). In the case of 2.5–4.0 vol.% O_2, no transition of the conversion curve happens until 300 s; this is because there is still a part of unoccupied Cu_2O surface which can react with O_2 directly. In the case of 8.0–21 vol.% O_2, the transition of the conversion curve occurs before 300 s when most part of the grain surface is covered by the product islands.

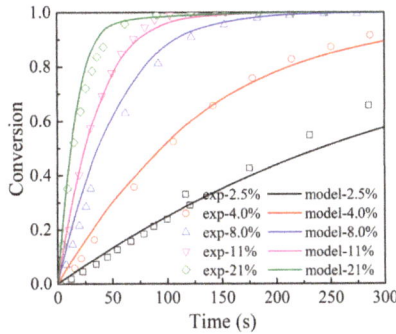

Figure 4. Effects of O_2 volume percentage (vol%) on conversion at 900 °C. ([O_2] = 2.5~21 vol%, N_2 as equilibrium gas. Dots: experimental data [42]; lines: model results.)

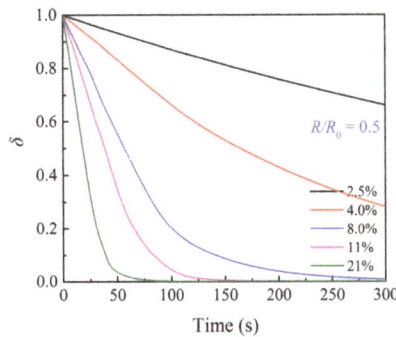

Figure 5. Profile of ratio of the unoccupied area on grain surface at $R/R_0 = 0.5$ during the oxidation of the particle in Figure 4.

3.2. Effects of Temperature

The experimental data of TGA obtained from the research of Clayton et al. [16] were used to validate the developed model and investigate the effects of temperature on the oxidation reaction behavior. In their study, two different Cu-based carriers, named as 50_TiO$_2$ material and 45_ZrO$_2$ material, were tested under different temperatures. The 50_TiO$_2$ material, supported by TiO$_2$, was prepared using the mechanical mixing method, and CuO loading capacity was 50 wt.%. The 45_ZrO$_2$ material, supported by ZrO$_2$/MgO, was prepared using the freeze granulation method, and CuO loading capacity was 45 wt.%. To minimize mass transfer effects, small particle size (~40 μm) and a shallow layer of particles were used.

The calculated results using the multiscale model were compared with the TGA experimental results of Clayton et al. [16], as shown in Figure 6. Figure 6a shows conversion curves of a 50_TiO$_2$ material particle under different temperatures (600–800 °C). Figure 6b shows conversion curves of a 45_ZrO$_2$ material particle under different temperatures (600–700 °C). It shows that the conversion curves obtained from the model calculation agree well with the experimental data. The reaction rate in the initial fast stage is higher when the temperature is higher, although the increased amplitude gradually decreases with the increase in temperature. When the reaction time reaches 120 s, the final conversion increases with the increasing temperature. Comparing the reaction rate in the second lower reaction stage in Figure 6a,b, the reaction rate of a 50_TiO$_2$ material particle is slower than that of the 45_ZrO$_2$ material particle.

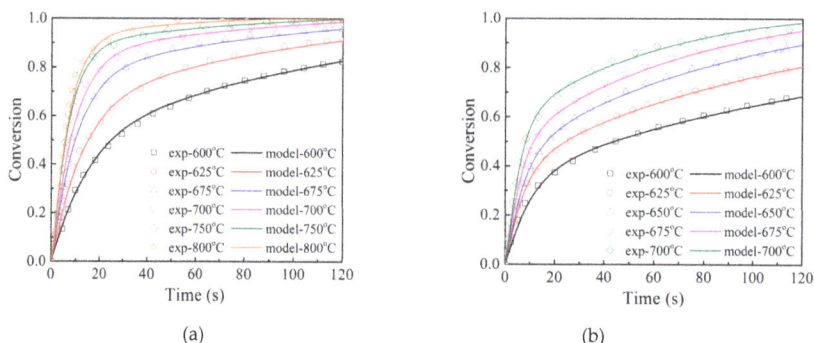

(a)

(b)

Figure 6. Effects of temperature on conversion of (**a**) 50_TiO$_2$ material and (**b**) 45_ZrO$_2$ material. ([O$_2$] = 21 vol%, [N$_2$] = 79 vol%. Dots: experimental data [16]; lines: calculated results.)

In the model calculation, there are three adjustable parameters included in the multiscale model: chemical reaction rate constant, k; the diffusivity of metal ions through the product layer, D_s; critical reactant layer thickness, h_c. These parameters can be written as a function of temperature:

$$k = k_0 \exp\left(-\frac{E_k}{R_g T}\right) \tag{23}$$

$$D_s = D_0 \exp\left(-\frac{E_D}{R_g T}\right) \tag{24}$$

$$h_c = aT + b \tag{25}$$

where k_0, E_k, D_0, E_D, a and b are kinetic parameters, and T is the temperature. The relation curves of these parameters obtained in the calculation of Figure 6 versus temperature were plotted, as shown in Figure 7. It can be seen from Equations (23)–(25) that the curves of ln$k$ versus $1/T$, lnD_s versus $1/T$, and h_c versus T are linear, and the values of kinetic parameters can be obtained from Figure 7. The kinetic parameters were calculated and shown in Table 1.

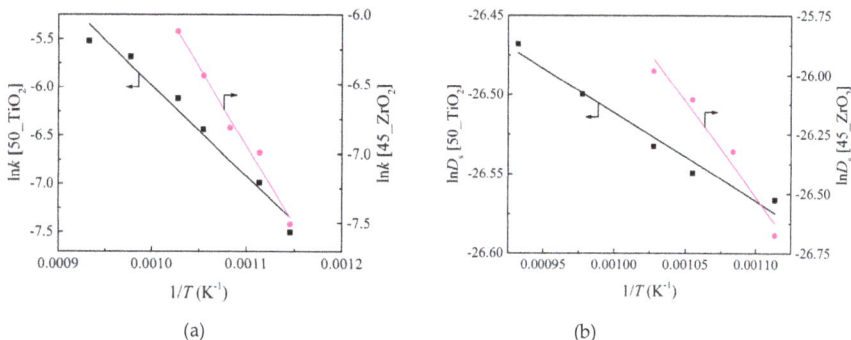

(a)

(b)

Figure 7. *Cont.*

(c)

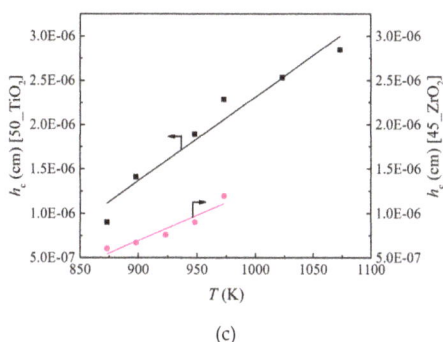

Figure 7. The relation curves of (**a**) ln$k$ versus $1/T$, (**b**) lnD_s versus $1/T$, and (**c**) h_c versus T.

Table 1. Values of the kinetic parameters for the oxidation of Cu-based oxygen carrier.

Parameter	k_0 (cm/s)	E_k (kJ/mol)	D_0 (cm^2/s)	E_D (kJ/mol)	a (cm/K)	b (cm)
50_TiO$_2$ material	29	78	5.3×10^{-12}	4.6	9.4×10^{-9}	-7.1×10^{-6}
45_ZrO$_2$ material	245	94	2.2×10^{-8}	67	5.7×10^{-9}	-4.4×10^{-6}

Figure 8 presents the modeling output of the profiles of the ratio of the unoccupied area on grain surface and local porosity during the 50_TiO$_2$_MM material oxidation reaction in Figure 6a. It shows an increase of the temperature will result in a faster reduction of the ratio of the unoccupied area on grain surface because the critical solid reactant layer thickness increases with the increasing temperature [37]. Figure 8b shows the local porosity profile under different temperatures. The solid product has a larger molar volume than the solid reactant, giving rise to a progressive decrease of porosity throughout the sorbent particle. It is clear that the porosity significantly decreases in the initial fast stage, followed by a slight change in the second stage, which is attributed to the fact that product islands finally cover the grain surface, as shown in Figure 8a, and hence the product layer diffusion fully controls the reaction process.

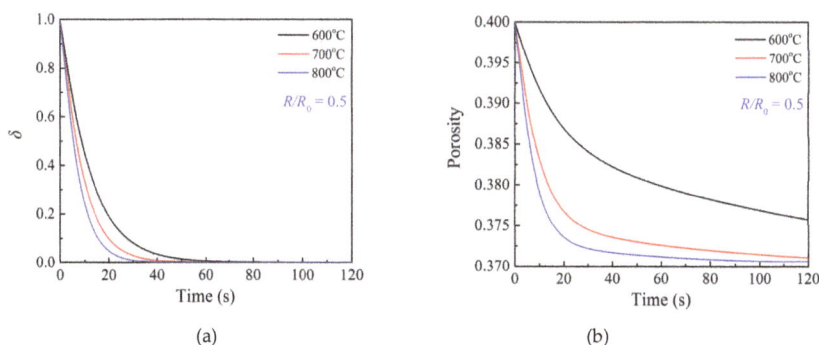

(a)

(b)

Figure 8. Profiles of (**a**) ratio of unoccupied area on grain surface at $R/R_0 = 0.5$ and (**b**) local porosity at $R/R_0 = 0.5$ during the 50_TiO$_2$_MM material oxidation reaction in Figure 6a.

3.3. Effects of Particle Structure

The particle structure has a significant influence on the oxidation reaction performance of an oxygen carrier with O$_2$, including particle porosity, particle size, and grain size. The effects of these factors on oxidation performance were investigated using the model developed in this study.

The particle porosity plays an important role in the gas-solid reaction, and it is a key parameter considered in the preparation of an oxygen carrier particle. Particles with high porosity allow the reactant gas to achieve active sites easily, leading to a high reaction rate and overall conversion in specific residence time. Figure 9 presents the modeling results of the effects of porosity on the oxidation reaction. It can be seen that when the initial particle porosity is larger, the gas concentration inside the particle is higher, and the overall conversion of the corresponding particle is higher. However, as shown in Figure 9b, there is no further change in the reaction rate of the particle when the initial particle porosity is larger than 0.4, where the internal gas diffusion resistance is small enough inside the particle, as shown in Figure 9c. Moreover, the duration time of the initial fast reaction stage is influenced by the initial particle porosity, which is attributed to the effects of the particle porosity on the distribution of oxygen concentration inside the particle, as described in Equation (9).

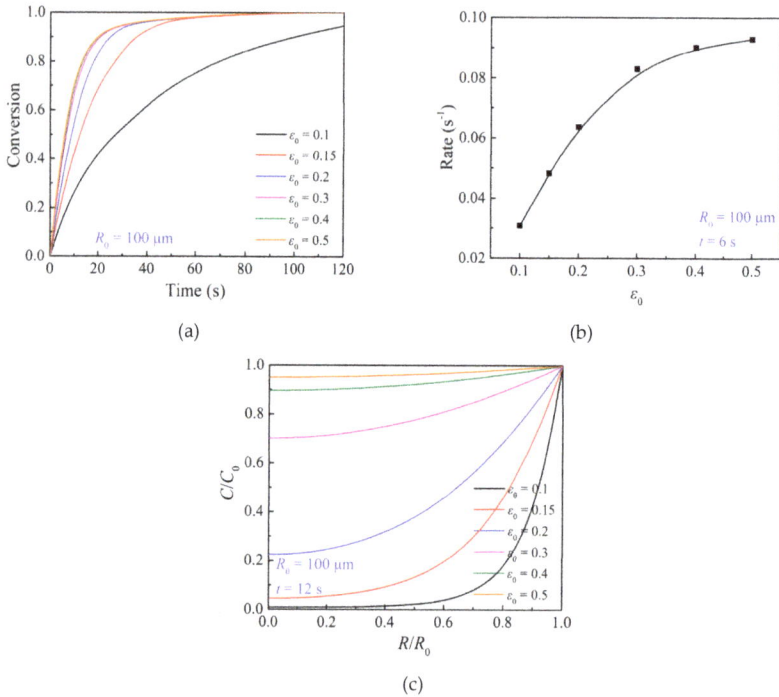

Figure 9. Effects of particle porosity on (**a**) the overall conversion of the particle, (**b**) reaction rate of the particle, and (**c**) oxygen concentration at $R/R_0 = 0.5$ during the oxidation reaction at 800 °C.

The particle size usually affects the external and internal transfer of O_2 significantly [8,44]. Figure 10 shows the modeling results of the effects of the particle size on the oxidation reaction. As can be seen, the decrease in the initial particle size will increase the reaction rate in the initial fast stage and the overall conversion of the particle. The corresponding oxygen concentration is higher with the decrease in the initial particle size. The radial distribution of oxygen concentration shows uniformity when the initial particle size is smaller than 100 μm, while there is a vast difference between the oxygen concentration at the particle center and that at the particle external surface when the initial particle size is larger than 300 μm. When the particle radius is smaller than 100 μm, a change in the particle size has no significant effect on the reaction performance, and the same conclusion was also reported in the research of García-Labiano [17]. This is because the internal gas diffusion resistance is small enough, as shown in Figure 10b, and there is a uniform reaction inside the particle, as also proposed by other researchers [45,46].

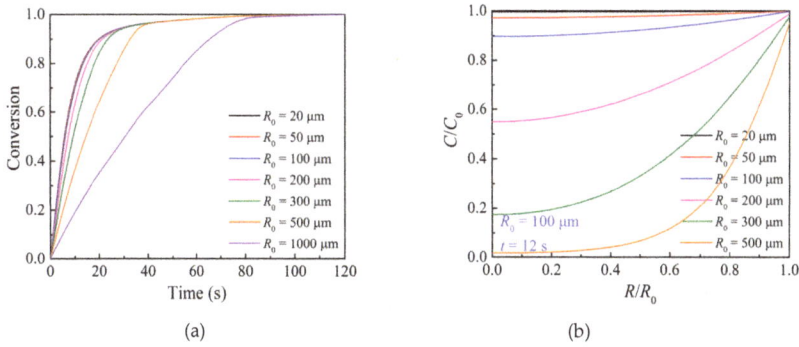

Figure 10. Effects of particle size on (**a**) the overall conversion of the particle and (**b**) oxygen concentration at $R/R_0 = 0.5$ during the oxidation reaction at 800 °C.

The grain size determines the reaction interface and affects the oxidation process significantly. Figure 11a illustrates the overall conversion variation with the change of grain size as oxidation reaction proceeds. As smaller grains could provide more reaction interface areas, a smaller grain size results in a higher overall conversion in the same reaction time period. As observed, the particle with 30 nm grains could achieve the overall conversion up to 1.0 at 40 s, larger than two times of that of the particle with 200 nm grains. Therefore, the small grain size is desirable for high reaction performance of the oxygen carrier. It should be noted that the smaller grain size will lead to a small pore size, which will increase the gas internal diffusion resistance, as shown in Figure 11b.

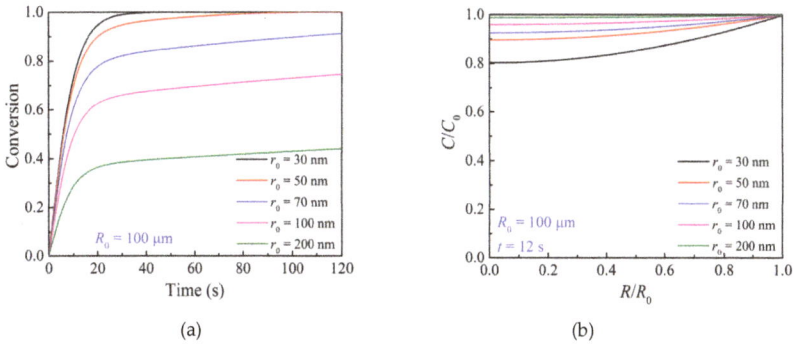

Figure 11. Effects of grain size on (**a**) the overall conversion of the particle and (**b**) oxygen concentration at $R/R_0 = 0.5$ during the oxidation reaction at 800 °C.

4. Discussion

4.1. External Mass Transfer

The external mass transfer was considered in the particle scale model, and it might have an influence on the oxygen concentration in the external surface of the particle. As described in Equation (17), during the three particle structure parameters, only particle size affects the external mass transfer coefficient, k_g. Correspondingly, it is shown in Figure 10b that the particle size has an effect on the oxygen concentration in the particle external surface, while Figures 9c and 11b show that the particle porosity and grain size have no effects on the oxygen concentration on the particle external surface. As observed in Figure 10b, even in the worst case simulated for the Cu_2O oxidation reaction, where the particle size is up to 1 mm, the oxygen concentration on the particle external surface only decreased by 5% with respect to the ambient gas concentration.

To analyze the effects of external mass transfer on the oxidation reaction behavior, the mass transfer rate of oxygen was analyzed. The mass transfer rate of oxygen to the particle is expressed as $4\pi R_0^2 k_g (C_0 - C_s)$. The reaction rate of the particle can be written as $4\pi R_0^2 k_{int}(C_s - C_e)$, where k_{int} is a comprehensive reaction rate constant describing the physical/chemical step inside the particle, including the internal gas diffusion, surface chemical reaction, and product layer diffusion. There should be $k_g(C_0 - C_s) = k_{int}(C_s - C_e)$. The magnitude of external mass transfer resistance can be described as $(1/k_g)/(1/k_{int})$, which is equal to $(C_0 - C_s)/(C_s - C_e)$. Therefore, the value of $(C_0 - C_s)/(C_s - C_e)$ was used to evaluate the importance of external mass transfer resistance during the oxidation reaction, as shown in Figure 12. It can be seen that even in the case of particle size of 1 mm, the importance of external mass transfer resistance is smaller than 0.05. Correspondingly, in the theoretical research of García-Labiano [17] and Sahir et al. [44], it was also predicted that the external mass transfer resistance hardly affected the reaction rate under the particle sizes normally used in a CLOU system, which was consistent with the calculation results in this work. In the experimental work of Chuang et al. [8], it was also proposed that the reaction of the particles (355–500 μm) was hardly controlled by external mass transfer. The external mass transfer resistance is small enough under the particle sizes investigated here, and it hardly affects the reaction performance. The oxidation behavior of the Cu_2O particle is controlled by the physical/chemical step inside the particle.

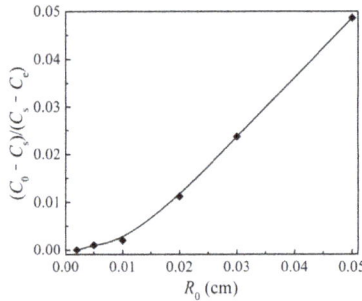

Figure 12. The importance of external mass transfer resistance described by $(C_0 - C_s)/(C_s - C_e)$ under different particle sizes at 800 °C.

4.2. Effectiveness Factor

The effectiveness factor, the ratio of the observed reaction rate to the intrinsic reaction rate, can be used to describe the relative importance of chemical reaction at grain scale versus internal gas diffusion inside the particle [47]. The chemical reaction step at grain scale includes the direct surface reaction on the unoccupied reactant surface and the indirect reaction on the solid product surface, where the reaction rate of the former step is very fast, and the reaction rate of the latter step is much slower.

The effectiveness factor, η, is expressed from a Thiele modulus, ϕ:

$$\eta = \frac{3}{\phi^2}(\phi\coth\phi - 1) \tag{26}$$

where

$$\phi = \frac{\int_0^{R_0} 4\pi R^2 \varphi \cdot dR}{\frac{4}{3}\pi R_0^3} \tag{27}$$

$$\varphi = R_0 \sqrt{\frac{\gamma}{D_e}} \tag{28}$$

$$\gamma = (1 - \delta)\frac{3\frac{r_1^2}{r_0^3}(1 - \varepsilon_0)k}{r_1(\frac{r_1}{r_2} - 1)\frac{k(C - C_e)}{D_s C_{si}} + 1} + \delta(1 - \frac{r_2^3}{r_0^3})\frac{(1 - \varepsilon_0)k}{h_c} \tag{29}$$

The value of the effectiveness factor changes during the reaction process. For the oxidation of the Cu_2O, the stoichiometric molar volume ratio of the solid product to the solid reactant is $Z = 1.05$, which is slightly larger than one. It means that there is just a slight decrease in the porosity inside the particle, as proved in Figure 8b. Therefore, the change of internal gas diffusion resistance during the reaction process is not obvious. The chemical reaction at grain scale is the main factor to affect the value of the effectiveness factor during the reaction process. Figure 13 shows the calculated results of the effectiveness factor and reaction rate against particle conversion under different particle sizes normally used in a CLOU system. As can be seen, under the particle radius of 50 µm, the effectiveness factor remains quite high throughout the entire reaction process, which means the whole oxidation reaction process is under the control of the chemical reaction at grain scale. The value of the effectiveness factor decreases with the increase in the particle size, which means the internal gas diffusion resistance is higher inside a particle with larger particle size, and the corresponding reaction rate decreases when the particle size increases. Under a certain particle size, as the reaction goes by, the value of the effectiveness factor increases, and the reaction rate decreases, which means chemical reaction resistance at grain scale increases. During the oxidation reaction process, the product islands grow quickly to cover the grain surface, leading to the reaction process changing from fast reaction stage to the second slower stage. Therefore, the increase in the chemical reaction resistance at grain scale results in the increase of the effectiveness factor.

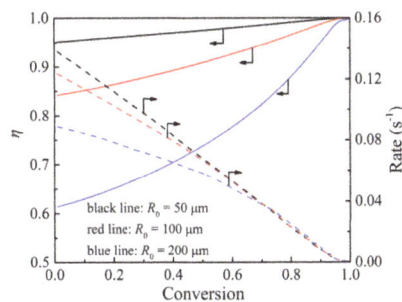

Figure 13. Calculated results of effectiveness factor and reaction rate against particle conversion under different particle sizes at 800 °C.

5. Conclusions

A multiscale model was established to describe the Cu-based oxygen carrier oxidation reaction with oxygen in this paper, including the surface scale, grain scale, and particle scale. The effects of oxygen partial pressure, temperature, and particle structure on the oxidation kinetics were studied using the developed model. The modeling results indicate that the transition of the conversion curve occurs when product islands cover most part of the grain surface. The oxygen partial pressure and particle structure have an obvious influence on the duration time of the fast reaction stage. An increase of the particle porosity, a decrease of the particle size, or a decrease of grain size will lead to better oxidation reaction behavior. However, no significant effect on the reaction was found when the initial particle porosity is larger than 0.4, or the initial particle radius is smaller than 100 µm, which is due to the negligible gas diffusion resistance. Furthermore, the importance of the external mass transfer and the effectiveness factor during the oxidation reaction process under the particle sizes normally used in a CLOU system were discussed to investigate the controlling step of the oxidation reaction of Cu-based oxygen carrier. It was concluded that the external mass transfer step hardly affects the reaction performance, and the oxidation behavior of the Cu_2O particle is controlled by the physical/chemical step inside the particle. The value of the effectiveness factor increases as the reaction goes by, which means the increased chemical reaction resistance at grain scale results from the growing number of product islands on the grain surface. Moreover, the multiscale model developed in this paper is expected to be useful for the oxidation kinetic analysis of other oxygen carriers used in CLOU.

Author Contributions: Conceptualization, H.W., Z.L. and N.C.; methodology, H.W. and Z.L.; software, H.W.; validation, H.W., Z.L. and N.C.; formal analysis, H.W.; investigation, H.W. and Z.L.; resources, Z.L. and N.C.; data curation, H.W. and Z.L.; Writing—Original Draft preparation, H.W.; Writing—Review and Editing, Z.L. and N.C.; visualization, Z.L.; supervision, N.C.; project administration, Z.L. and N.C.; funding acquisition, N.C.

Funding: This research was funded by National Natural Science Funds of China (Grant No. 51876095).

Conflicts of Interest: The authors declare no conflict of interest.

Nomenclature

C	concentration of O_2 inside the particle, mol/cm^3
C_0	concentration of O_2 in the ambient gas, mol/cm^3
C_e	equilibrium concentration of O_2, mol/cm^3
C_{met}	concentration of metal ions on the Cu_2O/CuO interface, mol/cm^3
C_s	concentration of O_2 on the particle external surface, mol/cm^3
D_1	diffusivity of gas through the particle pores, cm^2/s
D_e	effective diffusivity of gas through the particle pores, cm^2/s
D_K	Knudsen diffusivity, cm^2/s
D_{O_2}	O_2 molecular diffusivity, cm^2/s
D_s	diffusivity of metal ions through the product layer, cm^2/s
E_k, E_D	activation energy, kJ/mol
h_c	critical solid reactant layer thickness, cm
k	chemical reaction rate constant, cm/s
k_g	external mass transfer coefficient, cm/s
P	pressure, atm
P_e	equilibrium partial pressure of O_2, atm
r	grain radius, cm
r_0	initial grain radius, cm
r_1	grain radius of CuO/O_2 interface, cm
r_{1c}	critical grain radius of CuO/O_2 interface, cm
r_2	grain radius of Cu_2O/CuO interface, cm
r_{2c}	critical grain radius of Cu_2O/CuO interface, cm
R	particle radius, cm
R_0	initial particle radius, cm
R_g	the universal gas constant, $J/(mol \cdot K)$
t	reaction time, s
T	temperature, K
v	diffusion volume for molecule, cm^3
V_{CuO}^M	molar volume of CuO, cm^3/mol
$V_{Cu_2O}^M$	molar volume of Cu_2O, cm^3/mol
Z	stoichiometric molar volume ratio of the solid product to the solid reactant
α	local conversion
$\bar{\alpha}$	overall conversion
δ	ratio of the unoccupied area on the grain surface
ε	local porosity
ε_0	initial porosity
η	effectiveness factor
ϕ	Thiele modulus

References

1. Lyngfelt, A. Chemical-Looping Combustion of Solid Fuels—Status of Development. *Appl. Energy* **2014**, *113*, 1869–1873.
2. Lyngfelt, A.; Leckner, B.; Mattisson, T. A fluidized-bed combustion process with inherent CO_2 separation application of chemical-looping combustion. *Chem. Eng. Sci.* **2001**, *56*, 3101–3113.
3. Padurean, A.; Cormos, C.C.; Agachi, P.S. Pre-Combustion Carbon Dioxide Capture by Gas–Liquid Absorption for Integrated Gasification Combined Cycle Power Plants. *Int. J. Greenh. Gas Con.* **2012**, *7*, 1–11. [CrossRef]
4. Romeo, L.M.; Bolea, I.; Escosa, J.M. Integration of Power Plant and Amine Scrubbing to Reduce CO_2 Capture Costs. *Appl. Therm. Eng.* **2008**, *28*, 1039–1046.
5. Petrakopoulou, F.; Boyano, A.; Cabrera, M.; Tsatsaronis, G. Exergoeconomic and Exergoenvironmental Analyses of a Combined Cycle Power Plant with Chemical Looping Technology. *Int. J. Greenh. Gas Con.* **2011**, *5*, 475–482.
6. Mattisson, T.; Lyngfelt, A.; Leion, H. Chemical-looping with oxygen uncoupling for combustion of solid fuels. *Int. J. Greenh. Gas Con.* **2009**, *3*, 11–19. [CrossRef]
7. Abad, A.; Mattisson, T.; Lyngfelt, A.; Rydén, M. Chemical looping combustion in a 300 W_{th} continuously operating reactor system using a manganese-based oxygen carrier. *Fuel* **2006**, *85*, 1174–1185. [CrossRef]
8. Chuang, S.Y.; Dennis, J.S.; Hayhurst, A.N.; Scott, S.A. Kinetics of the oxidation of a co-precipitated mixture of Cu and Al_2O_3 by O_2 for chemical-looping combustion. *Energy Fuel* **2010**, *24*, 3917–3927.
9. Chuang, S.; Dennis, J.S.; Hayhurst, A.N.; Scott, S.A. Development and performance of Cu-based oxygen carriers for chemical-looping combustion. *Combust. Flame* **2008**, *154*, 109–121.
10. Adanez-Rubio, I.; Abad, A.; Gayán, P.; Diego, L.F.D.; García-Labiano, F.; Adánez, J. Identification of operational regions in the chemical-looping with oxygen uncoupling (CLOU)process with a Cu-based oxygen carrier. *Fuel* **2012**, *102*, 634–645.
11. Diego, L.F.D.; García-Labiano, F.; Adánez, J.; Gayán, P.; Abad, A.; Corbella, B.M.; Palaciosb, J.M. Development of Cu-based oxygen carriers for chemical-looping combustion. *Fuel* **2004**, *83*, 1749–1757.
12. Goldstein, E.A.; Mitchell, R.E. Chemical kinetics of copper oxide reduction with CO. *Proc. Combust. Inst.* **2011**, *33*, 2803–2810. [CrossRef]
13. Gayán, P.; Adánez-Rubio, I.; Abad, A.; Diego, L.F.D.; García-Labiano, F.; Adánez, J. Development of Cu-based oxygen carriers for chemical-looping with oxygen uncoupling (CLOU) process. *Fuel* **2012**, *96*, 226–238. [CrossRef]
14. Adánez-Rubio, I.; Arjmand, M.; Leion, H.; Gayán, P.; Abad, A.; Mattisson, T.; Lyngfelt, A. Investigation of combined supports for Cu-based oxygen carriers for chemical-looping with oxygen uncoupling (CLOU). *Energy Fuel* **2013**, *27*, 3918–3927. [CrossRef]
15. Diego, L.F.D.; Gayán, P.; García-Labiano, F.; Celaya, J.; Abad, A.; Adánez, J. Impregnated CuO/Al_2O_3 oxygen carriers for chemical-looping combustion, avoiding fluidized bed agglomeration. *Energy Fuel* **2005**, *19*, 1850–1856. [CrossRef]
16. Clayton, C.K.; Sohn, H.Y.; Whitty, K.J. Oxidation Kinetics of Cu_2O in Oxygen Carriers for Chemical Looping with Oxygen Uncoupling. *Ind. Eng. Chem. Res.* **2014**, *53*, 2976–2986. [CrossRef]
17. García-Labiano, F.; Diego, L.F.D.; Adánez, J.; Abad, A.; Gayán, P. Temperature variations in the oxygen carrier particles during their reduction and oxidation in a chemical-looping combustion system. *Chem. Eng. Sci.* **2005**, *60*, 851–862. [CrossRef]
18. Maya, J.C.; Chejne, F. Modeling of Oxidation and Reduction of a Copper-Based Oxygen Carrier. *Energy Fuel* **2014**, *28*, 5434–5444. [CrossRef]
19. Dennis, J.S.; Pacciani, R. The rate and extent of uptake of CO_2 by a synthetic, CaO-containing sorbent. *Chem. Eng. Sci.* **2009**, *64*, 2147–2157. [CrossRef]
20. Liu, W.; Dennis, J.S.; Sultan, D.S.; Redfern, S.A.T.; Scott, S.A. An investigation of the kinetics of CO_2 uptake by a synthetic calcium based sorbent. *Chem. Eng. Sci.* **2012**, *69*, 644–658. [CrossRef]
21. Szekely, J.; Evans, J.W.; Sohn, H.Y. *Gas–Solid Reactions*; Academic Press: New York, NY, USA, 1976.
22. Gattinoni, C.; Michaelides, A. Atomistic details of oxide surfaces and surface oxidation: The example of copper and its oxides. *Surf. Sci. Rep.* **2015**, *70*, 424–447.

23. Zhang, R.; Liu, H.; Zheng, H.; Ling, L.; Wang, B. Adsorption and dissociation of O_2 on the $Cu_2O(111)$ surface: Thermochemistry, reaction barrier. *Appl. Surf. Sci.* **2011**, *257*, 4787–4794.

24. Yu, X.; Zhang, X.; Tian, X.; Wang, S.; Feng, G. Density functional theory calculations on oxygen adsorption on the Cu_2O surfaces. *Appl. Surf. Sci.* **2015**, *324*, 53–60. [CrossRef]

25. Bao, J.; Li, Z.; Sun, H.; Cai, N. Experiment and rate equation modeling of Fe oxidation kinetics in chemical looping combustion. *Combust. Flame* **2013**, *160*, 808–817.

26. Li, Z.S.; Fang, F.; Cai, N.S. Characteristic of solid product layer of $MgSO_4$ in the reaction of MgO with SO_2. *Sci. China Technol. Sci.* **2010**, *53*, 1869–1876. [CrossRef]

27. Fang, F.; Li, Z.S.; Cai, N.S.; Tang, X.Y.; Yang, H.T. AFM investigation of solid product layer of $MgSO_4$ generated on MgO surfaces in the reaction of MgO with SO_2 and O_2. *Chem. Eng. Sci.* **2011**, *66*, 1142–1149. [CrossRef]

28. Tang, X.Y.; Li, Z.S.; Fang, F.; Cai, N.S.; Yang, H.T. AFM investigation of the morphology of $CaSO_4$ product layer formed during direct sulfation on polished single-crystal $CaCO_3$ surfaces at high CO_2 concentrations. *Proc. Combust. Inst.* **2011**, *33*, 2683–2689.

29. Schulz, K.H.; Cox, D.F. Photoemission and low-energy-electron-diffraction study of clean and oxygen-dosed Cu_2O (111) and (100) surfaces. *Phys. Rev. B* **1991**, *43*, 1610–1621.

30. Le, D.; Stolbov, S.; Rahman, T.S. Reactivity of the Cu_2O (100) surface: Insights from first principles calculations. *Surf. Sci.* **2009**, *603*, 1637–1645.

31. Zhou, L.J.; Zou, Y.C.; Zhao, J.; Wang, P.P.; Feng, L.L.; Sun, L.W.; Wang, D.J.; Li, G.D. Facile synthesis of highly stable and porous Cu_2O/CuO cubes with enhanced gas sensing properties. *Sens. Actuators B* **2013**, *188*, 533–539. [CrossRef]

32. Trinh, T.T.; Tu, N.H.; Le, H.H.; Ryu, K.Y.; Le, K.B.; Pillai, K.; Yi, J. Improving the ethanol sensing of ZnO nano-particle thin films—The correlation between the grain size and the sensing mechanism. *Sens. Actuators B* **2011**, *152*, 73–81.

33. Umar, A.; Alshahrani, A.A.; Algarni, H.; Kumar, R. CuO nanosheets as potential scaffolds for gas sensing applications. *Sens. Actuators B* **2017**, *250*, 24–31. [CrossRef]

34. Wang, M.; Liu, J.; Shen, F.; Cheng, H.; Dai, J.; Long, Y. Theoretical study of stability and reaction mechanism of CuO supported on ZrO_2 during chemical looping combustion. *Appl. Surf. Sci.* **2016**, *367*, 485–492.

35. Alvarez, D.; Abanades, J.C. Determination of the critical product layer thickness in the reaction of CaO with CO_2. *Ind. Eng. Chem. Res.* **2005**, *44*, 5608–5615.

36. Marks, L.D. Modified wulff constructions for twinned particles. *J. Cryst. Growth* **1983**, *61*, 556–566. [CrossRef]

37. Li, Z.S.; Sun, H.M.; Cai, N.S. Rate Equation Theory for the Carbonation Reaction of CaO with CO_2. *Energy Fuel* **2012**, *26*, 4607–4616.

38. Wang, H.; Li, Z.S.; Fan, X.X.; Cai, N.S. Rate-Equation-Based Grain Model for the Carbonation of CaO with CO_2. *Energy Fuel* **2017**, *31*, 14018–14032. [CrossRef]

39. Rapp, R.A. The high temperature oxidation of metals forming cation-diffusing scales. *Metall. Mater. Trans. A* **1984**, *15*, 765–782. [CrossRef]

40. Fuller, E.N.; Ensley, K.; Giddings, J.C. New method for prediction of binary gas-phase diffusion coefficients. *Ind. Eng. Chem. Res.* **1996**, *58*, 19–27. [CrossRef]

41. Khoshandam, B.; Kumar, R.V.; Allahgholi, L. Mathematical modeling of co2removal using carbonation with CaO: The grain model. *Korean J. Chem. Eng.* **2010**, *27*, 766–776. [CrossRef]

42. Adánez-Rubio, I.; Gayán, P.; Abad, A.; García-Labiano, F.; De Diego, L.F.; Adánez, J. Kinetic analysis of a cu-based oxygen carrier: Relevance of temperature and oxygen partial pressure on reduction and oxidation reactions rates in chemical looping with oxygen uncoupling (CLOU). *Chem. Eng. J.* **2014**, *256*, 69–84.

43. Whitty, K.; Clayton, C. Measurement and Modeling of Kinetics for Copper-Based Chemical Looping with Oxygen Uncoupling. In Proceedings of the 2nd International Conference on Chemical Looping, Darmstadt, Germany, 26–28 September 2012.

44. Sahir, A.H.; Lighty, J.A.S.; Sohn, H.Y. Kinetics of Copper Oxidation in the Air Reactor of a Chemical Looping Combustion System using the Law of Additive Reaction Times. *Ind. Eng. Chem. Res.* **2011**, *50*, 566–580. [CrossRef]

45. Alonso, M.; Criado, Y.A.; Abanades, J.C.; Grasa, G. Undesired effects in the determination of CO$_2$ carrying capacities of CaO during TG testing. *Fuel* **2014**, *127*, 52–61.
46. Sedghkerdar, M.H.; Mostafavi, E.; Mahinpey, N. Investigation of the Kinetics of Carbonation Reaction with Cao-Based Sorbents Using Experiments and Aspen Plus Simulation. *Chem. Eng. Commun.* **2015**, *202*, 746–755.
47. Yao, J.G.; Zhang, Z.; Sceats, M.; Maitland, G.C.; Fennell, P.S. Two-phase fluidized bed model for pressurized carbonation kinetics of calcium oxide. *Energy Fuel* **2017**, *31*, 11181–11193. [CrossRef]

materials

MDPI

Article

Numerical Assessment on Rotation Effect of the Stagnation Surface on Nanoparticle Deposition in Flame Synthesis

Lilin Hu, Zhu Miao, Yang Zhang, Hai Zhang * and Hairui Yang

Key Laboratory for Thermal Science and Power Engineering of Ministry of Education,
Tsinghua University-University of Waterloo Joint Research Center for Micro/Nano Energy & Environment
Technology Department of Energy and Power Engineering, Tsinghua University, Beijing 100084, China;
hll17@mails.tsinghua.edu.cn (L.H.); oceanskymz@126.com (Z.M.); yang-zhang@tsinghua.edu.cn (Y.Z.);
yhr@tsinghua.edu.cn (H.Y.)
* Correspondence: haizhang@tsinghua.edu.cn; Tel.: +86-10-62773153

Received: 27 March 2019; Accepted: 17 April 2019; Published: 26 April 2019

Abstract: The effect of rotation of the stagnation surface on the nanoparticle deposition in the flame stabilizing on a rotating surface (FSRS) configuration was numerically assessed using CFD method. The deposition properties including particle trajectories, deposition time, temperature and surrounding O_2 concentration between the flame and stagnation surface were examined. The results revealed that although flame position is insensitive to the surface rotation, the temperature and velocity fields are remarkably affected, and the deposition properties become asymmetric along the burner centerline when the surface rotates at a fast speed (rotational speed $\omega \geq 300$ rpm). Particles moving on the windward side have similar deposition properties when the surface rotates slowly, but the off-center particles on the leeward side have remarkable longer deposition time, lower deposition temperature, and lower surrounding O_2 concentration, and they even never deposit on the surface when the surface rotates at a high speed. The rotation effect of the stagnation surface can be quantitatively described by an analogous Karlovitz number (Ka'), which is defined as the ratio of characteristic residence time of moving surface to the aerodynamics time induced by flame stretch. For high quality semiconducting metal oxide (SMO) films, it is suggested that $Ka' \geq 1$ should be kept.

Keywords: flame synthesis; flame stabilizing on a rotating surface (FSRS); rotational speed; particle deposition; Karlovitz number

---

1. Introduction

Nano-sized semiconducting metal oxide (SMO) materials such as TiO_2, SnO_2, and ZnO are widely used in photocatalysis, gas sensors and solar cells [1–3]. A few techniques have been developed to fabricate nanoparticles, such as the sol-gel method [4], co-precipitation method [5], hydrothermal method [6], impregnation method [7,8], colloidal method [9], and flame synthesis method [10,11]. Among them, the flame synthesis method has a great potential for massive production due to its merits of high throughput, simple post treatment and relatively low cost [12,13]. During the synthesis, the precursors doped in the fuel-oxidizer mixtures undergo rapid decomposition and oxidation in a high temperature flame zone, and the vapor-phase metal oxides turn into fine particles through nucleation, collision and sintering in the post flame zone [14,15]. Clearly, the temperature and velocity distributions in the post flame zone are of great significance for nanoparticle size, uniformity, and deposition on the film.

To well control the thickness and quality of SMO films in a single-step gas-to-film deposition process, flame stabilizing on a rotating surface (FSRS) method was proposed by Wang et al. [16]. This method uses an aerodynamic nozzle to generate a laminar premixed flat flame opposing to a film substrate. The substrate is affixed with a rotating disk which is cooled by the ambient air or cooling water [17,18]. Due to the large temperature gradient between the flame sheet and the cold solid surface, a strong thermophoretic force is induced, driving synthesized particles to deposit on the substrate to form a SMO film within a few milliseconds. Previous studies [16,17,19–21] found that FSRS method can well control flame temperature, particle deposition time and gas composition, so it is effective to obtain the desired crystal phase of the nanoparticles, and fabricate sensing films with high sensitivity, selectivity, and stability performance.

However, the previous studies are mostly done with a specific or a narrow range of the rotational speed of the stagnation surface. The rotation effect of the stagnation surface or the film substrate is scarcely assessed. In some studies [16,19], flame position and shape were even assumed to be barely affected by the rotating stagnation surface. In fact, it is straightforward that when the stagnation surface rotates very fast, the ambient cold gas could be entrained into the space between the flame sheet and the stagnation surface. If so, the particle deposition time, local temperature and O_2 concentration could be affected, resulting in the different size and phase of the synthesized particles [16]. To properly set up the operational parameter for FSRS flame synthesis process, it is necessary to assess the influence of the rotational speed.

Therefore, in this paper, 3-D CFD simulations on the stagnation flow with FSRS setting, especially in the post flame zone, are conducted at different rotational speed (ω) in a range of 0 to 600 rpm (round per minute). Based on the simulated velocity and temperature fields, the effects of rotational speed of the stagnation surface on deposition process, including the deposition time, temperature and O_2 concentration are assessed. A guide to select a proper ω for the rotating stagnation surface is to be provided. The study is also helpful to understand the effect of rotating surface on the stagnation flames.

2. Numerical Methods

Figure 1 is schematic diagram of the experimental system using FSRS method [18]. A nozzle is placed above a rotating disk. When the combustible mixture is lit, a flame is stabilized between the nozzle exit and the rotating disk, and the top surface of the disk becomes a stagnation surface. The position of the flame depends on the fuel properties and the aerodynamic stretch induced by the imposing flow. For sensor fabrication, a set of substrates are mounted on the solid surface and right below the nozzle exit.

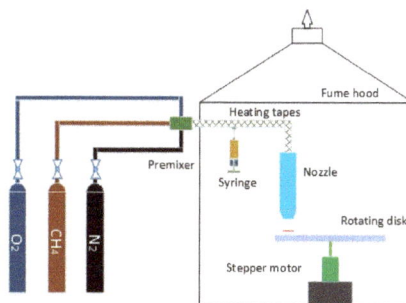

Figure 1. Schematic diagram of the experimental system using flame stabilizing on a rotating surface (FSRS) method [18].

The computational domain is shown in Figure 2. Similar to the configuration used in the previous study [18], the centerline of burner offsets 120 mm from that of the disk, which spins at ω rpm. The distance between the burner exit and the stagnation surface is 30 mm. The burner exit has an inner diameter of 10 mm. Inert coflow (Ar) is supplied from the external circular outlet with inner and outer diameters of 11 mm and 14 mm respectively. Namely, the thickness of nozzle at exit is 0.5 mm, and thus the edge of the nozzle exit is assumed to be infinitely thin in CFD simulation. The overall dimension of the computational domain is 304 mm in diameter and 35 mm in height. The organometallic precursors-doped premixed mixtures are injected from the fixed burner, stabilizing a premixed flame above the rotating stagnation surface. Several substrates are placed on the stagnation surface. The horizontal dimension of the substrate is 10 mm × 10 mm.

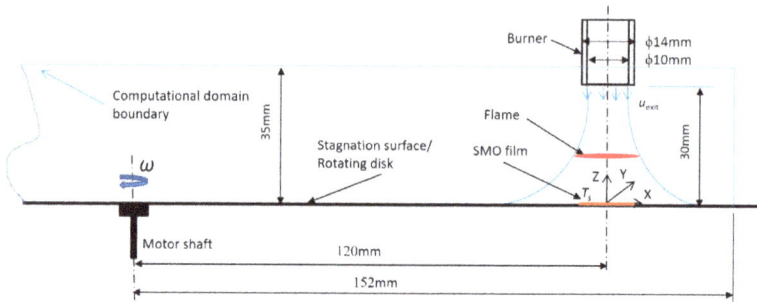

Figure 2. Schematic of the computational domain for the FSRS simulation.

As shown in Figure 2, the Z axis denotes the direction along the axis of the burner. The X and Y axis denote the radial and tangential direction of the rotating disk. In Z direction, the bottom region includes the thin gas layer between the rotating surface and the flame, and thus exponential meshing is adopted. The rest region includes the jet flow and surrounding environment, and uniform meshing is employed. In X- and Y-directions, the domain is divided into nine sectors. Denser gridding is used in the sectors close to the nozzle centerline. The mesh size increases with the distance from the nozzle. Before the simulations, the grid-independency test is performed. The temperature and velocity profiles along the centerline of the flame are obtained when the number of mesh is 500,000, 700,000, 900,000, and 1,300,000, as shown in Figure 3. It can be seen that for mesh number higher than 900,000, the temperature and velocity profiles barely change with mesh number. Therefore, the total mesh number is set to ~1 M. The meshing is presented in Figure 4.

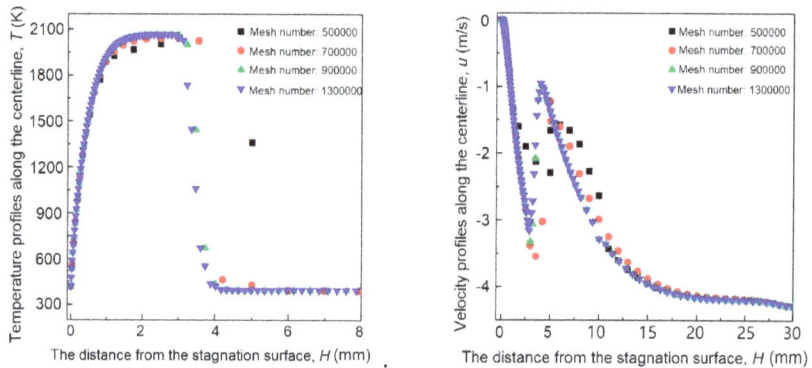

Figure 3. Mesh independence test: (**a**) Temperature profiles along the centerline of the flame at different mesh number; (**b**) velocity profiles along the centerline of the flame at different mesh number.

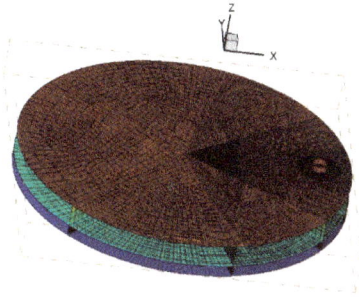

Figure 4. Mesh of the computational domain.

Consistent with the reported experiments [19], simulations are conducted for the lean premixed $C_2H_4/O_2/Ar$ flames (3.9% C_2H_4-29.5% O_2-Ar, equivalence ratio $\phi = 0.4$) at an initial temperature of 393 K. The velocity at the nozzle exit is 4.29 m/s and 5.52 m/s for the premixed reacting gas and co-flow gas (Ar), respectively. The Ar flow is injected to prevent the impact of surrounding air to the deposition process. Correspondingly, Reynolds number is estimated as 1700 and 2070 and the mean strain rate is 143 s^{-1} and 174 s^{-1}.

The temperature of the stagnation surface T_s ($Z = 0$) will change with ω. According to the previous experimental study [16], the surface temperature, T_s is different at different ω's, and their relationship can be expressed by an empirical correlation T_s (K) = 464 − 0.15 ω, in which the unit of ω is rpm (round per minute).

A modified 3-step global mechanism [22] is adopted to describe the chemistry of $C_2H_4/O_2/Ar$ mixtures. The global mechanism is integrated into Fluent in Chemkin format. The thermal and transport parameters of the species are retrieved from USC-Mech II files [23]. Viscous model is selected for laminar simulation, and thermal diffusion is considered. The burner exits are set as velocity boundaries with a constant temperature of 393 K. The other boundaries are set as pressure outlet and the ambient gas is air. The bottom boundary is set as non-slip wall boundary rotating around Z-axis. Uniform velocity distribution is set for the exits of the premixed unburned gas and the coflow gas. Steady and pressure-based solver is used.

Particle movements are also studied in the simulation. The central flame section is selected for analyses along the burner centerline. Based on the classical theory, thermophoretic velocity V_r of particles can be calculated by Equation (1) [24], in which T is the local temperature, K; v is the local gas kinematic viscosity, m^2/s; and a is momentum accommodation coefficient to describe the momentum exchanges during particle collision and is normally set as 0.9 [25].

$$V_r = \frac{-3v \cdot \nabla T}{(4 + 0.5\pi a)T} \tag{1}$$

The particle velocity is the summation of fluid velocity and thermophoretic velocity. With the particle velocity, the particle path from the flame sheet to the stagnation surface is determined.

As shown in Figure 5, the temperature distributions on X-Z plane are symmetric along the nozzle axis even when $\omega = 600$ rpm, indicating that the rotational speed has slight influence on X-Z plane. Given that the rotation of the stagnation surface may have the greatest influence on the tangential direction, the Y-Z slice (40 mm width × 5mm height) along the flame centerline is selected. The windward side of the flame refers to the area where $Y \leq 0$ and the leeward side is the area where $Y > 0$ according to the rotation direction of the surface.

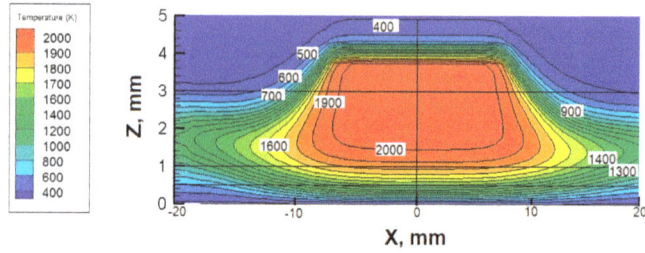

Figure 5. Temperature contours at $\omega = 600$ rpm in X-Z plane (unit: K).

3. Results and Discussions

3.1. Axial Velocity Contours at Different Rotational Speeds

Figure 6 shows the axial velocity contours adjacent to the flame and the stagnation surface at different ω's. When the disk is in stationary ($\omega = 0$), axial velocity distribution is symmetric along the centerline of the nozzle. When the disk rotates ($\omega > 0$), the velocity field on the windward side is pushed upwards. The phenomenon is more obvious as ω increases. When $\omega = 600$ rpm, the influencing zone covers nearly the entire space between the flame and stagnation surface. While $\omega \leq 300$ rpm, the influencing zone is limited to the space near the edge of the flame on the windward side. However, the central area under the flame, i.e., the main synthesis zone is barely affected. Since the horizontal area of the substrate of the SMO film is usually smaller than 10 mm × 10 mm, the influence of the surface rotation is minor when $\omega \leq 300$ rpm, validating the assumption used in the previous study [19].

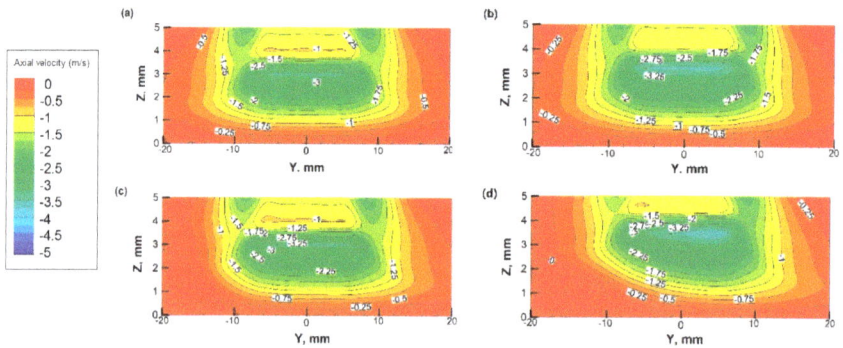

Figure 6. Axial velocity contours at different rotational speed in Y-Z plane (unit: m/s): (**a**) $\omega = 0$; (**b**) $\omega = 100$ rpm; (**c**) $\omega = 300$ rpm; (**d**) $\omega = 600$ rpm.

3.2. Temperature Fields at Different Rotational Speeds

Figure 7 shows the temperature contours in the central section adjacent to the stagnation surface at different ω's.

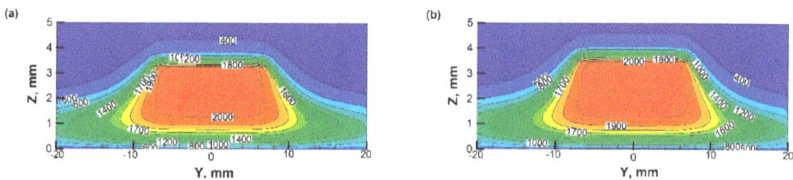

Figure 7. *Cont.*

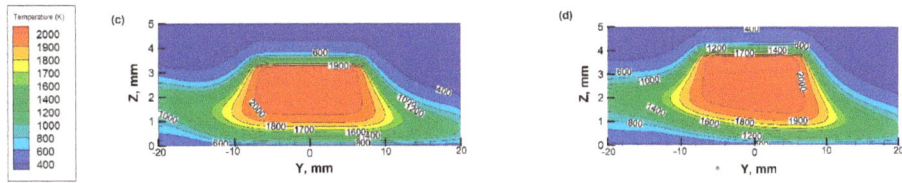

Figure 7. Temperature contours at different rotational speed in *Y-Z* plane (unit: K): (**a**) $\omega = 0$; (**b**) $\omega = 100$ rpm; (**c**) $\omega = 300$ rpm; (**d**) $\omega = 600$ rpm.

It can be seen that the flame is stabilized ~3 mm above the stagnation surface, with a diameter of ~30 mm. When the surface is in stationary or rotates slowly (e.g., $\omega \leq 100$ rpm), the temperature distribution in the synthesis space is symmetric along the centerline of the nozzle. However, when $\omega = 600$ rpm, the temperature fields are very asymmetric, as some cold gas is induced into the bottom of the flame on the windward side. Again, the temperature field results indicate the flame is insignificantly affected by the rotating surface when $\omega \leq 300$ rpm. Based on the temperature and velocity distribution, it is clear that the influence of the rotating stagnation surface is mainly concentrated in the near wall area where the deposition of nanoparticles occurs. While the nucleation and growth of particles near the flame front remain almost unaffected. Therefore, the deposition process is particularly studied.

3.3. Particle Deposition Trajectory at Different Rotational Speeds

To exam the rotation effect on the deposition process, the deposition trajectories, time, temperature, and mean surrounding O_2 concentration of 17 selected particles in particle deposition zone are computed at different ω's. The above temperature distributions show that the temperature on the plane 2 mm above the stagnation surface have very small difference at different ω's. Since the horizontal dimension of the substrate of film is about 10 mm × 10 mm, the initial positions of the tracked particles are set on the plane 2 mm above the stagnation surface, shown as the solid dots in Figure 8. The initial radial position of the particles y_0 locates at 0, ±0.5 mm, ±1.0 mm, ±1.5 mm, ±2.0 mm, ±2.5 mm, ±3.0 mm, ±4.0 mm and ±5.0 mm respectively.

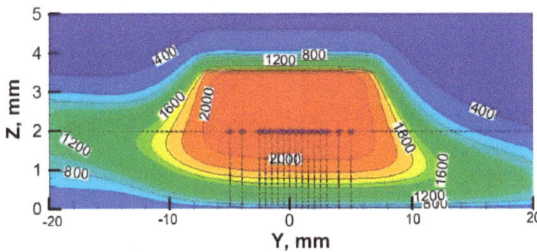

Figure 8. Schematic of the initial settings for tracked particles (The case with $\omega = 300$ rpm).

Figure 9 depicts the deposition paths of the tracked particles when $\omega = 0$ and $\omega = 300$ rpm respectively, both at the same surface temperature ($T_s = 419$ K). As expected, when $\omega = 0$, particle deposition paths are symmetrically distributed along the central axis. Due to the thermophoretic force, the tracked particles deposit within a circle of $\phi 40$ mm on the stagnation surface. When the thermophoretic force is arbitrarily excluded, the tracked particles move with the gas flow and none of them deposits on the film. Under the condition with $\omega = 300$ rpm, some particle paths become asymmetric, with the trajectories shifting to leeward side. The tracked particles away from the centerline on the leeward side deposit outside the circle of $\phi 40$ mm. While on the windward side, some particles are entrained into the upward flow generated by thin gas layer of the moving stagnation surface and the flame jet flow. As a result, some particles cannot even approach the wall surface. The closer to the stagnation surface, the greater the impact is. Consequently, particle deposition efficiency decreases.

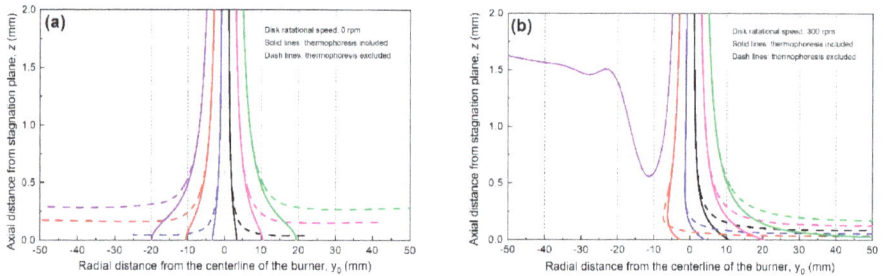

Figure 9. Particle trajectories of the tracked particles when the stagnation surface is (**a**) in stationary and (**b**) ω = 300 rpm (at the same surface temperature).

3.4. Particle Deposition Time at Different Rotational Speeds

Deposition time of the tracked particle τ_d is defined as the time that particle experiences from the initial tracked position (near the flame front) to its deposition location on the stagnation surface. Figure 10a shows the variation of τ_d with T_s when T_s = 374 – 456 K, corresponding to the T_s's when surface rotates in the range of 50–600 rpm. Clearly, τ_d is insensitive to T_s. It only decreases a little at high T_s because of large thermophoretic force and remains nearly constant for particles impinging from the central flame surface (d ≤ 5 mm). Out of the central flame surface, τ_d increases rapidly due to the weak thermophoretic force.

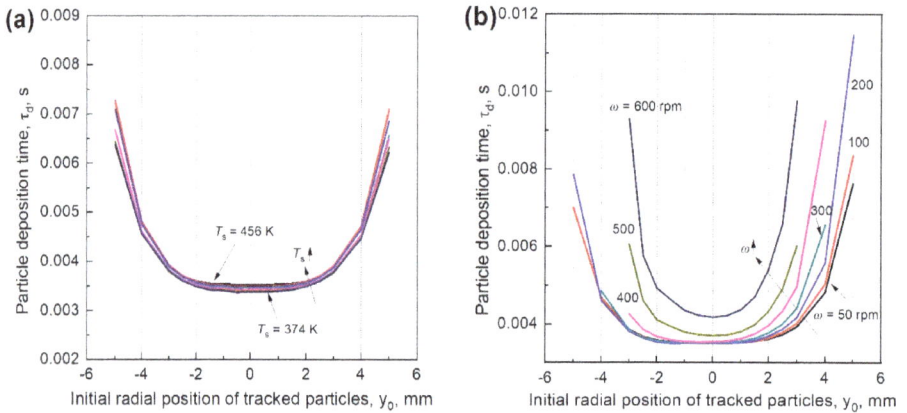

Figure 10. Deposition time of the tracked particles when the stagnation surface is (**a**) in stationary and (**b**) rotates at different speeds.

When the surface rotates, shown in Figure 10b, τ_d remarkably changes with ω. With the increase of ω, τ_d increases and the particles with constant τ_d are limited to a smaller area. This phenomenon can be attributed to the entrained cold gas around the moving stagnation surface, leading to a smaller temperature gradient in the near wall region, thus a weaker thermophoretic force. When $\omega \leq 300$ rpm, the particles right above the solid surface still have the approximate τ_d. For the particles out of the deposition area, τ_d is more sensitive to ω on the leeward side. When $\omega > 300$ rpm, τ_d increases rapidly with ω. For the particles whose initial positions are not in the flame center, the variation trend is more obvious. Consistent with the trajectory results, particles further away from the flame center has infinite

τ_d. The results show that when $\omega \le 300$ rpm, the particles deposit on the stagnation surface have nearly the same τ_d, which is conducive for high-quality and uniform film products.

3.5. Particle Deposition Temperature at Different Rotational Speeds

Since the nanoparticles are very small, their temperatures are assumed to be equal to the fluid temperature. Particle deposition temperature, T_d, is the average particle temperature during the deposition process. As shown in Figure 11, T_d is insensitive to T_s. When $\omega = 0$, T_d in the central flame region is relatively uniform, while T_d far away from the flame center is much smaller. When $\omega > 0$, T_d becomes smaller for all tracked particles as some low temperature ambient gas is entrained into the flame region. At a large ω, especially when $\omega > 300$ rpm, T_d with different initial positions are of great difference. On the windward side, only the particles around the flame center can deposit on the surface, and they mostly have rather high T_d. However, slightly off the centerline on the leeward side, e.g., when $y_0 = 1$ mm, T_d remarkably decreases from the peak value, and this is because particles on the leeward side are blown away from the flame center zone and deposit in a low temperature zone as shown in Figure 8. In the center area with $\phi 20$ mm, the effect of the rotating surface on T_d is minor when $\omega \le 300$ rpm.

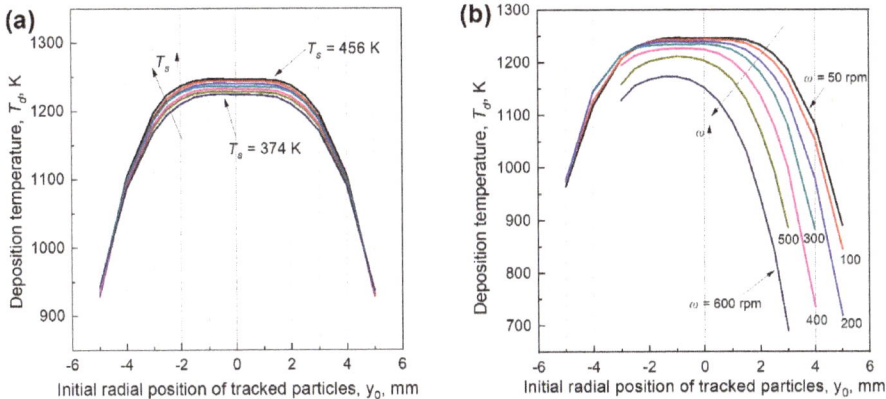

Figure 11. Deposition temperature of the tracked particles when the stagnation surface is (a) in stationary and (b) rotates at different speeds.

3.6. Particle Deposition O_2 Concentration at Different Rotational Speeds

The averaged O_2 mole fraction X_{O2} in the deposition zone between the flame and surface is important for the product phase [26]. Figure 12 show the X_{O2} for the tracked particles at different T_s's varying from 374 K to 456 K. When $\omega = 0$, X_{O2} is insensitive to T_s. For the particles moving down to the surface from flame within the area of $\phi 5$ mm, X_{O2} is rather constant, while particles from the edge of the flame experience lower X_{O2}. The further are the particles away from the flame center, the lower X_{O2} is. When the surface rotates at a slow speed, e.g., $\omega \le 300$ rpm, X_{O2} for the tracked particles are approximately the same near the center of the flame, while it decreases remarkably for the particles moving down on the leeward side. When $\omega > 300$ rpm, X_{O2} of all tracked particles decreases with ω. The decrease rate is higher when the particles are located further away from the center or at a higher ω. In addition, the influence on the leeward particles is more significant than the windward side.

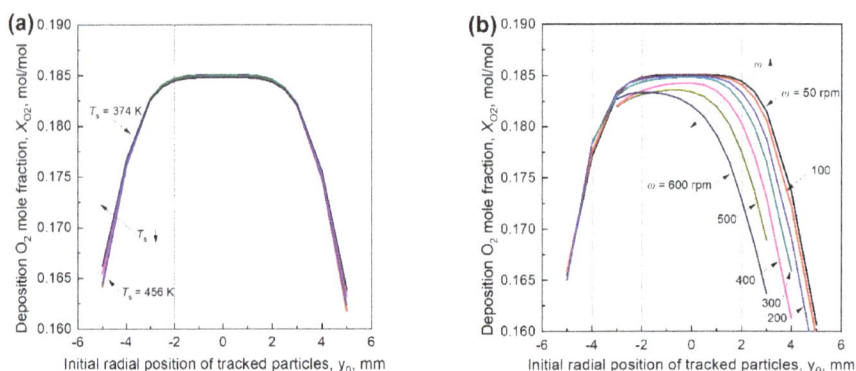

Figure 12. Averaged O_2 concentration around the tracked particles when the stagnation surface is (a) in stationary and (b) rotates at different speeds.

3.7. The Effect of the Tangential Velocity and Flame Stretch Rate

The tangential velocity $u_{surf} = \omega \cdot r$, in which r is the distance of the film to the axis of the disk, is more reasonable to be used to evaluate the impact of rotational stagnation surface on deposition process since it includes the radial position of the film substrate. In Figure 13a, the average deviation Δ represents the difference caused by disk rotation. It can be seen that increasing the tangential velocity (rotational speed) has the most profound influence on the deposition time τ_d, following by the deposition temperature T_d and the local O_2 concentration.

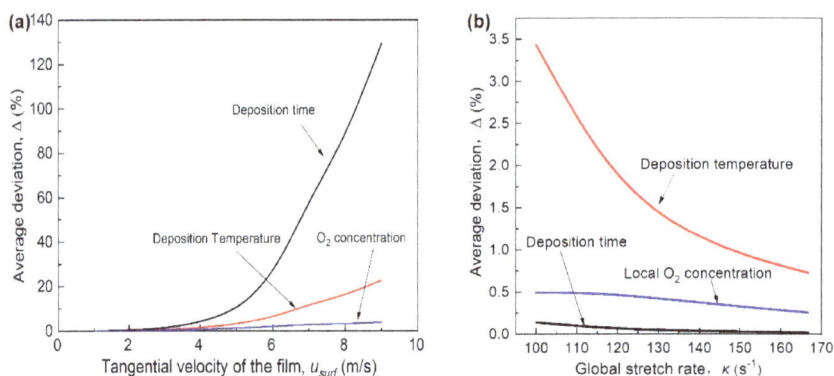

Figure 13. The effects of (a) tangential velocity and (b) stretch rate on the nanoparticle deposition.

The flame stretch is a dominant factor to the flow and temperature fields for the stagnation flame. Thus, for FSRS flame synthesis process, the flame stretch should be another key influencing factor besides the rotational speed, or tangential velocity. In the stagnation configuration, the intensity of global flame stretch can be expressed as: $k = u_{exit} / L$, in which u_{exit} is the exit velocity of the flow and L is the distance between the nozzle and the solid surface.

Figure 13b shows the variation of the deposition performance caused by the flame stretch, based on the simulation results at different u_{exit}'s with $\omega = 300$ rpm. It can be seen the increasing of k will reduce the deviation of τ_d, T_d and O_2 concentration in the deposition zone, opposite to the increasing of tangential velocity. The stretch rate effect is weak on τ_d and O_2 concentration, but still significant on T_d.

Indeed, the rotation of stagnation surface induces a characteristic residence time in the horizontal direction for the deposition zone, and this characteristic time can be expressed as $\tau_{surf} = D_f / u_{surf}$, where D_f is the diameter of the flame front, approximate to the nozzle diameter. The flame stretching introduces a characteristic residence time in vertical direction. This characteristic time can be expressed as $\tau_{fl} = 1 / k$. Combining the two effects, we can quantify the total effects by introducing an analogous non-dimensional Karlovitz number (Ka') [27,28], which is defined as the ratio of characteristic residence time induced by the moving surface to the aerodynamic time induced by flame stretch.

$$Ka' = \frac{\tau_{surf}}{\tau_{fl}} = D_f / u_{surf} \cdot k \qquad (2)$$

Figure 14 shows the variation of the average deviations of τ_d, T_d and O_2 mole fraction with Ka'. It can be seen that the deviations for these deposition properties decrease rapidly with increasing Ka', and becomes less than 10% when $Ka' \geq 1$. Basically, the deposition O_2 concentration and T_d are more weakly affected by Ka', and τ_d is greatly affected by Ka'. For high quality SMO films, it is suggested that $Ka' \geq 1$ should be kept.

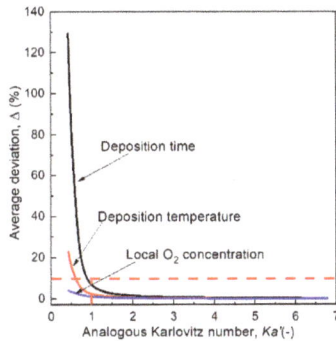

Figure 14. Relation between the average deviation of deposition parameters and analogous Karlovitz number.

4. Conclusions

FSRS (flame stabilizing on a rotating surface) is a proved method with a single step deposition for nano SMO film fabrication. With well controlled flame temperature, particle deposition time and gas composition, it is effective to obtain the desired crystal phase of the nanoparticles, and fabricate sensing films with high sensitivity, selectivity, and stability performance. However, the temperature and velocity fields for the nano particle deposition could be significantly influenced by the rotation of the stagnation surface. In this paper, 3-D CFD simulation was conducted to assess the effect of rotating surface on the nanoparticles deposition in the FSRS configuration for the premixed C_2H_4/air flames. It was found that although flame position is insensitive to the rotation of stagnation surface, the temperature and velocity fields could be remarkably affected. When the surface rotates slowly, the temperature and velocity fields near the flame barely change. When the surface rotates at a fast speed (e.g., $\omega > 300$ rpm), the flame on the windward side tilts upward and the entire flame moves to the leeward side. Based on the simulated results, the deposition trajectories, time, temperature, and mean surrounding O_2 concentration of selected particles between the flame and the surface are computed at different surface rotational speeds. When the surface is in stationary, the deposition of the particles is caused by the thermophoretic force and symmetric along the nozzle centerline. Those deposition properties for the particles moving from the flame center are insensitive to the surface temperature variation, which could be caused by the surface rotation. The deposition properties for the particles from the flame center zone are rather close. When the surface rotates slowly ($\omega \leq 300$ rpm for the present configuration), the particles moving from flame center on the windward side have similar deposition properties. The particles from the flame but in off-center area on the leeward side have remarkable longer deposition time, lower deposition temperature and lower surrounding O_2 concentration. When the surface rotates faster, the changes in deposition properties are severer. The effect of rotational surface can be described by analogous Karlovitz number (Ka'), which is defined as the ratio of characteristic residence time of moving surface to the aerodynamics time induced by flame stretch. With proper settings of the operational parameters of FSRS method, the negative impacts caused by the rotation of the stagnation surface could be minimized. Based on the simulation results of this paper, for high quality SMO films, it is suggested that $Ka' \geq 1$ should be satisfied.

Author Contributions: L.H. is main investigator; Z.M. is another main investigator; Y.Z. is a co-advisor, H.Z. is the supervisor and H.Y. reviewed and provided guidance.

Funding: This work was supported by the Natural Science Foundation of China (Project No. 51706119 and 51476088).

Acknowledgments: The advices given by Hai Wang at Stanford University are highly appreciated.

Conflicts of Interest: The authors declare no conflict of interest.

References

1. Kim, Y.; Yoon, Y.; Shin, D. Fabrication of Sn/SnO$_2$ composite powder for anode of lithium ion battery by aerosol flame deposition. *J. Anal. Appl. Pyrolysis* **2009**, *85*, 557–560. [CrossRef]
2. Jensen, J.R.; Johannessen, T.; Wedel, S.; Livbjerg, H. A study of Cu/ZnO/Al$_2$O$_3$ methanol catalysts prepared by flame combustion synthesis. *J. Catalysis* **2003**, *218*, 67–77. [CrossRef]
3. Akpan, U.G.; Hameed, B.H. Parameters affecting the photocatalytic degradation of dyes using TiO$_2$-based photocatalysts: A review. *J. Hazard. Mater.* **2009**, *170*, 520–529. [CrossRef] [PubMed]
4. Kulal, A.B.; Dongare, M.K.; Umbarkar, S.B. Sol-gel synthesized WO$_3$ nanoparticles supported on mesoporous silica for liquid phase nitration of aromatics. *Appl. Catal. B Environ.* **2016**, *182*, 142–152. [CrossRef]
5. Habibi, M.H.; Mardani, M. Co-precipitation synthesis of nano-composites consists of zinc and tin oxides coatings on glass with enhanced photocatalytic activity on degradation of reactive blue 160 KE2B. *Spectrochim. Acta A* **2015**, *137*, 785–789. [CrossRef]
6. Liu, J.L.; Fan, L.Z.; Qu, X.H. Low temperature hydrothermal synthesis of nano-sized manganese oxide for supercapacitors. *Electrochim. Acta* **2012**, *66*, 302–305. [CrossRef]
7. Feng, D.Y.; Rui, Z.B.; Lu, Y.B.; Ji, H.B. A simple method to decorate TiO$_2$ nanotube arrays with controllable quantity of metal nanoparticles. *Chem. Eng. J.* **2012**, *179*, 363–371. [CrossRef]
8. Lu, Y.B.; Wang, J.M.; Yu, L.; Kovarik, L.; Zhang, X.W.; Hoffman, A.S.; Gallo, A.; Bare, S.R.; Sokaras, D.; Kroll, T.; et al. Identification of the active complex for CO oxidation over single-atom Ir-on-MgAl$_2$O$_4$ catalysts. *Nat. Catal.* **2019**, *2*, 149–156. [CrossRef]
9. Mozaffari, S.; Li, W.H.; Thompson, C.; Ivanov, S.; Seifert, S.; Lee, B.; Kovarik, L.; Karim, A.M. Colloidal nanoparticle size control: Experimental and kinetic modeling investigation of the ligand-metal binding role in controlling the nucleation and growth kinetics. *Nanoscale* **2017**, *9*, 13772–13785. [CrossRef]
10. Zhang, Y.Y.; Li, S.Q.; Deng, S.L.; Yao, Q.; Tse, D.S. Direct synthesis of nanostructured TiO$_2$ films with controlled morphologies by stagnation swirl flames. *J. Aerosol Sci.* **2012**, *44*, 71–82. [CrossRef]
11. Wooldridge, M.S. Gas-phase combustion synthesis of particles. *Prog. Energy Comb. Sci.* **1998**, *24*, 63–87. [CrossRef]
12. Pratsinis, S.E. Flame aerosol synthesis of ceramic powders. *Prog. Energy Comb. Sci.* **1998**, *24*, 197–219. [CrossRef]
13. Yue, R.L.; Dong, M.; Ni, Y.; Jia, Y.; Liu, G.; Yang, J.; Liu, H.D.; Wu, X.F.; Chen, Y.F. One-step flame synthesis of hydrophobic silica nanoparticles. *Powder Technol.* **2013**, *235*, 909–913. [CrossRef]
14. Memarzadeh, S.; Tolmachoff, E.D.; Phares, D.J.; Wang, H. Properties of nanocrystalline TiO$_2$ synthesized in premixed flames stabilized on a rotating surface. *Proc. Combust. Inst.* **2011**, *33*, 1917–1924. [CrossRef]
15. Gutsch, A.; Mühlenweg, H.; Krämer, M. Tailor-made nanoparticles via gas-phase synthesis. *Small* **2005**, *1*, 30–46. [CrossRef] [PubMed]

16. Tolmachoff, E.D.; Abid, A.D.; Phares, D.J.; Campbell, C.S.; Wang, H. Proc. Synthesis of nano-phase TiO_2 crystalline films over premixed stagnation flames. *Combust. Inst.* **2009**, *32*, 1839–1845. [CrossRef]

17. Nikraz, S.; Wang, H. Dye sensitized solar cells prepared by flames stabilized on a rotating surface. *Proc. Combust. Inst.* **2013**, *34*, 2171–2178. [CrossRef]

18. Miao, Z.; Zhang, Y.; Zhang, H.; Liu, Q.; Yang, H.R. Experimental study on gas sensing properties of TiO_2 film produced by flame synthesis. In Proceedings of the China National Symposium on Combustion, Xi'an, China, 31 October–2 November 2014.

19. Tolmachoff, E.D.; Memarzadeh, S.; Wang, H. Nanoporous titania gas sensor films prepared in a premixed stagnation flame. *J. Phys. Chem. C* **2011**, *115*, 21620–21628. [CrossRef]

20. Mädler, L.; Roessler, A.; Pratsinis, S.E.; Sahm, T.; Gurlo, A.; Barsan, N.; Weimar, U. Direct formation of highly porous gas-sensing films by in situ thermophoretic deposition of flame-made Pt/SnO_2 nanoparticles. *Sensor Actuat. B-Chem.* **2006**, *114*, 283–295. [CrossRef]

21. Wang, Y.; He, Y.M.; Lai, Q.H.; Fan, M.H. Review of the progress in preparing nano TiO_2: An important environmental engineering material. *J. Environ. Sci.* **2014**, *26*, 2139–2177. [CrossRef] [PubMed]

22. Westbrook, C.K.; Dryer, F.L. Simplified Reaction Mechanisms for the Oxidation of Hydrocarbon Fuel in Flames. *Comb. Sci. Tech.* **1981**, *27*, 31–43. [CrossRef]

23. Wang, H.; You, X.; Joshi, A.V.; Davis, S.G.; Laskin, A.; Egolfopoulos, F.N.; Law, C.K. High-Temperature Combustion Reaction Model of H2/CO/C1-C4 Compounds. Available online: http://ignis.usc.edu/USC_Mech_II.htm (accessed on 18 April 2019).

24. Waldmann, L.; Schmitt, K.H. Thermophoresis and diffusiophoresis of aerosol. In *Aerosol Science*; Academic Press: London, UK, 1966; pp. 137–162.

25. Friedlander, S.K. *Smoke, Dust and Haze: Fundamentals of Aerosol Dynamics*, 2nd ed.; Oxford University Press: New York, NY, USA, 2000.

26. Zhao, B.; Uchikawa, K.; Mccormick, J.R.; Ni, C.Y.; Chen, J.G.; Wang, H. Ultrafine anatase TiO_2 nanoparticles produced in premixed ethylene stagnation flame at 1 atm. *Proc. Combust. Inst.* **2005**, *30*, 2569–2576. [CrossRef]

27. Law, C.K. *Combustion Physics*; Cambridge University Press: Cambridge, UK, 2006.

28. Taamallah, S.; Shanbhogue, S.J.; Ghoniem, A.F. Turbulent flame stabilization modes in premixed swirl combustion: Physical mechanism and Karlovitz number-based criterion. *Combust. Flame* **2016**, *166*, 19–33. [CrossRef]

materials

MDPI

Article

Effects of the Limestone Particle Size on the Sulfation Reactivity at Low SO$_2$ Concentrations Using a LC-TGA

Runxia Cai [1], Yiqun Huang [1], Yiran Li [1], Yuxin Wu [1], Hai Zhang [1,2], Man Zhang [1,*], Hairui Yang [1] and Junfu Lyu [1]

[1] Key Laboratory for Thermal Science and Power Engineering of Ministry of Education, Department of Energy and Power Engineering, Tsinghua University, Haidian District, Beijing 100084, China; cairx14@mails.tsinghua.edu.cn (R.C.); huangyq1993@163.com (Y.H.); li-yr15@mails.tsinghua.edu.cn (Y.L.); wuyx09@mail.tsinghua.edu.cn (Y.W.); haizhang@mail.tsinghua.edu.cn (H.Z.); yhr@mail.tsinghua.edu.cn (H.Y.); lvjf@mail.tsinghua.edu.cn (J.L.)

[2] Tsinghua University-University of Waterloo Joint Research Center for Micro/Nano Energy and Environment Technology, Tsinghua University, Haidian District, Beijing 100084, China

* Correspondence: zhangman@mail.tsinghua.edu.cn; Tel.: +86-10-6277-3384

Received: 29 March 2019; Accepted: 1 May 2019; Published: 8 May 2019

Abstract: Limestone particle size has a crucial influence on SO$_2$ capture efficiency, however there are few studies on the sulfation reactivity, which covers a broad range of particle sizes at low SO$_2$ concentrations. In this paper, a large-capacity thermogravimetric analyzer (LC-TGA) was developed to obtain the sulfur removal reaction rate under a wide range of particle sizes (3 μm–600 μm) and SO$_2$ concentrations (250 ppm–2000 ppm), and then compared with the results of a traditional fixed bed reactor and a commercial TGA. The experimental results showed that the LC-TGA can well eliminate the external mass transfer and obtain a better measurement performance. Both the final conversion and the reaction rate reduced with the decreasing of SO$_2$ concentration, but ultrafine limestone particles still showed the good sulfation reactivity even at 250 ppm SO$_2$. An empirical sulfation model was established based on the experimental results, which can well predict the sulfation process of different limestone particle sizes at low SO$_2$ concentrations. The model parameters have a strong negative correlation against the particle size, and the fit of the reaction order of SO$_2$ was found to be about 0.6. The model form is very simple to incorporate it into available fluidized bed combustion models to predict SO$_2$ emission.

Keywords: Limestone; particle size; sulfation; TGA; model

---

1. Introduction

Limestone is widely used for SO$_2$ capture in circulating fluidized bed (CFB) boilers for its low price and availability [1,2]. Under air combustion conditions, limestone involves first calcination to the porous CaO, and then the reaction with sulfur containing gas to CaSO$_4$ [3]. As the molar volume of CaSO$_4$ is about three times larger than that of CaO [4], pore filling or pore plugging on the surface of a CaO particle will block the further reaction in the inner core. Thus, one of the main drawbacks of desulfurization by limestone is the low utilization rate of calcium [5].

Many factors will influence the sulfation process and the maximum calcium utilization rate, such as temperature, limestone properties, steam, particles size, SO$_2$ concentration and so on [6]. SO$_2$ capture is strongly affected by the temperature. High temperature will cause the sintering of sorbent particles and thermodynamic instability of CaSO$_4$ under reducing conditions, and low temperature will reduce the calcination rate and inhibit the pore development, thus the maximum sulfur capture efficiency in atmospheric fluidized beds is usually achieved at 850 °C or a little lower [7–10]. Limestones vary greatly in properties, and the geological properties also have strong influence on the reactivity of CaO. The consensus view is that older limestones tend to be more compact and less reactive than younger limestones [6]. Steam can also affect calcination, sintering and sulfation reactions of the limestone, and a small amount of water vapor may have positive effect on the calcium conversion rate [11].

Limestone particle size also has a crucial influence on SO_2 capture efficiency. Small pores result in a high reaction rate but will be easily plugged during the sulfation process [12], so only the superficial surface layer participates in the reaction [13]. Since this superficial area of a sorbent particle increases directly with the decreasing of particle size, the reaction rate of smaller limestone is much higher than that of larger ones. Due to the formation of the product layer, even after exposure to SO_2 gas for several hours, a considerable amount of CaO in the core area still remained unreacted for coarse particles [14]. Therefore, it is commonly believed that smaller particles can achieve a faster reaction rate and a higher calcium conversion [15]. However, some researchers noted that under actual CFB conditions, the residence time of very fine particles was restricted and could not meet the requirement of the contact time for SO_2 capture [16,17]. Thus, it was considered that the optimum sorbent particle size should be close to the circulating ash for a longer residence time and a relatively higher reaction rate. Nevertheless, recent industrial practices have found that fine and even ultrafine (<10 μm) limestones can realize a high SO_2 capture efficiency with a low Ca/S molar ratio [2,18], so it is meaningful to investigate the optimum sorbent particle size under different conditions. Among all the influencing factors for the particle size optimization, the reactivity at low SO_2 concentrations is most significant.

A brief summary of investigations of a broad range of particle sizes under atmospheric conditions is listed in Table 1. Although many researchers have investigated the effects of the limestone particle size on the reaction between SO_2 and CaO, there are just a few research results of a certain type of limestone whose particle size ranges from several micrometers to several hundred micrometers. Combining the experimental data from different scholars can be a possible way to solve this issue, but the differences of the limestone properties and the calcination conditions may cause some deviation in model prediction [19,20]. Adánez et al. [21] compared three different structural sulfation models and found that the same model parameters could not predict the conversion curves of different particle sizes of sorbents. Modification of model parameters with respect to the limestone particle size should be introduced for better prediction. Based on the ideas of the shrinking core model, Obras-Loscertales et al. [22] proposed a two-step sulfation model, which can predict the sulfation conversion of particles between 200 μm and 630 μm with similar parameters. However, the fitting thickness of the product layer, which was about 30 μm, could not be suitable for the particles smaller than 60 μm, otherwise the product layer will be thicker than the sorbent particle. To establish the sulfation model which is valid for different particle sizes, it is very important to first obtain the sulfation conversion curves covering the range of particle sizes from several micrometers to several hundred micrometers.

Additionally, both the reaction rate and the final sulfation conversion will strongly be affected by SO_2 concentration. Most experiments are undertaken in a relatively high SO_2 concentration over 1000 ppm, but the SO_2 emission standards for CFB boilers in most countries are normally lower than 100 ppm [23]. What is worse, with the implement of the updated national emission regulation in China, SO_2 emission is required to be not higher than 35 mg/Nm^3 (~12 ppm) [24,25]. Most models regard the CaO sulfation as a first-order reaction [19,26], but Borgwardt et al. [20] obtained the apparent reaction orders of 0.62 ± 0.07. To obtain a better prediction at low SO_2 concentrations, it is worthy to investigate the desulfurization characteristics at different SO_2 concentrations, especially at the low SO_2 concentrations.

Table 1. A brief summary of investigations of a broad range of particle sizes under atmospheric conditions.

Author	Method	Particle Size (μm)	Temperature (°C)	Atmosphere	SO$_2$ Concentration (ppm)
Simons [27]	Fixed Bed Reactor	1–78	850	$N_2 + O_2 + CO_2 + H_2O + SO_2$	297–315
Zarkanitis [28]	TGA	53–350	700–850	$N_2 + O_2 + SO_2$	3000–5000
Milne [29]	Dispersed-Phase Isothermal Reactor	4.1–49	980–1171	$N_2 + O_2 + SO_2$	1480
Adánez [21,26]	TGA Bubbling Bed Reactor	158–1788	800–900	$N_2 + O_2 + SO_2 + CO_2$	2500
Mattisson [19]	Fixed Bed Reactor	45–2000	850	$N_2 + O_2 + SO_2 + CO_2$	1500
Fan [15]	Differential Bed Reactor	7.5–150 (modified)	900	$N_2 + O_2 + SO_2$	3900
Laursen [30]	Fixed Bed	212–355	850	$N_2 + O_2 + SO_2$	2250
Abanades [31]	TGA	70–1000	850	$N_2 + O_2 + SO_2$	500–5000
Obras-Loscertales [22]	TGA Bubbling Bed Reactor	200–630	800–950	$N_2 + O_2 + SO_2 + CO_2$	1500–4500

Thermogravimetric analyzers (TGAs), fixed bed reactors and bubbling bed reactors are commonly used to study the effects of particle size on sulfation reaction. Although the heat and mass transfer conditions in a bubbling bed reactor are rather similar to that in a CFB boiler, it is not suitable for fine particles because of the particle escape. A fixed bed reactor can easily overcome this issue, but SO$_2$ concentrations at the reactor outlet are always lower than the main stream, so differential conditions cannot be achieved at the beginning of the reaction. In addition, measurement of the gas component is also limited by the response and accuracy of the instrument. The TGA has satisfactory repeatability and accuracy, but small crucibles will lead to the particle packing, which will restrict the reaction due to the external mass transfer.

Therefore, in this paper, a large-capacity TGA (LC-TGA) was developed to investigate the effects of limestone particle size on the sulfation reactivity at low SO$_2$ concentrations. Furthermore, an empirical model was proposed based on the experimental data.

2. Materials and Methods

The schematics of the LC-TGA is shown in Figure 1. Gaseous mixture through the mixture chamber was introduced from the top of the quartz tube to react with limestone samples on a quartz crucible (Shengfan Shiying Corporation, Lianyungang City, Jiangsu Province, China), whose diameter was 32 mm. The quartz tube was fixed with the heating furnace (Yuzhi Mechanical and Electrical Corporation, Shanghai, China), which can slide along the rail at the highest velocity of 10 mm/s. A K-type thermocouple, whose measurement accuracy is ±0.4%, was placed above the quartz crucible to record the reaction temperature. Mass variation of limestone samples was automatically recorded by a MT-WKX204 analytical balance produced by Mettler Toledo (Zurich, Switzerland). The maximum weight is over 100 g, and the readability can be down to 0.1 mg. The analytical balance was installed inside a water-cooled jacket, which was fixed on the ground to ensure stable readings. Inert Ar atmosphere was introduced into the water-cooled jacket to create a positive pressure environment, so the balance can be protected from the high temperature corrosion problems. A rubber seal O ring was installed between the quartz tube and the water-cooled jacket to prevent gas leakage, thus the exhaust gas was the mixture of the reaction gas and the protecting gas.

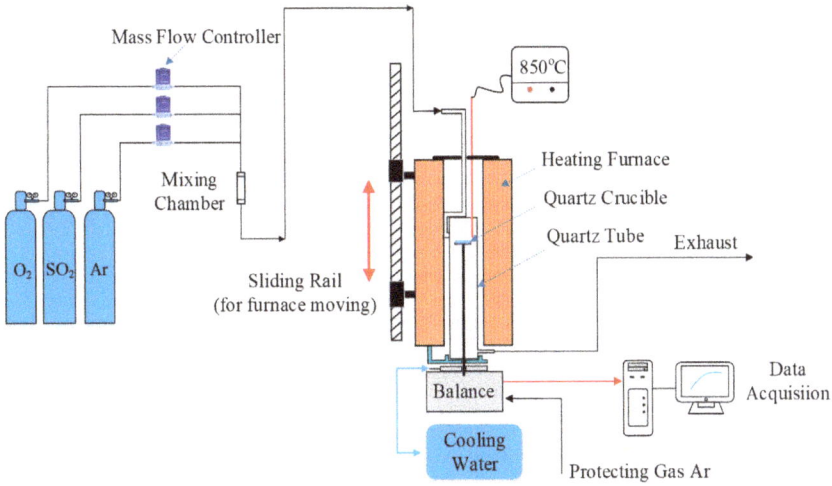

Figure 1. The schematic of the large-capacity thermogravimetric analyzer (LC-TGA) system.

Two types of limestone samples from China and Korea were sieved by an ultrasonic sieving machine (Sanyuantang Mechanical Corporation, Xinxiang County, Henan Province, China) into seven groups with narrow cuts, including 0–20 µm, 20–38 µm, 38–75 µm, 75–106 µm, 106–200 µm, 200–400 µm and 400–600 µm. The particle size distributions measured by Malvern are shown in Figure 2. The main cut sizes of the seven groups are listed in Table 2. The X-Ray Fluorescence (XRF) analysis for each size cut of limestone was performed, and the relative deviation of CaO content was smaller than 2%. Thus, the component was assumed the same for each type of limestone, and the average values are listed in Table 3.

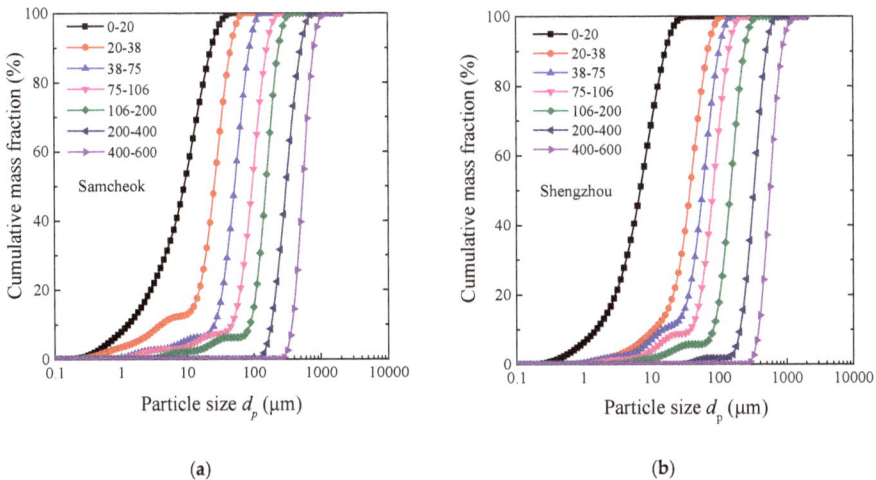

Figure 2. The particle size distributions of two limestone samples with narrow cut. (**a**) Korea Samcheok; (**b**) China Shengzhou.

Table 2. The main cut sizes of the seven groups of the two kinds of limestone samples.

Particle Size (µm)	Korea Samcheok				China Shengzhou			
	d_{10}	d_{50}	d_{90}	d_{32}	d_{10}	d_{50}	d_{90}	d_{32}
0–23	1.4	10.3	26.7	3.4	1.7	7.8	18.4	3.5
20–38	4.5	29.1	51.9	7.4	10.8	37.9	69.7	17.9
38–75	29.1	58.2	99.3	26.3	16.7	65.4	114.0	26.5
75–106	52.0	106.5	180.7	35.5	42.8	93.6	158.0	39.4
106–200	98.2	172.6	271.3	86.2	93.5	165.4	265.1	91.9
200–400	205.8	346.2	577.9	320.4	228.1	362.4	560.0	308.1
400–600	421.4	609.7	886.1	586.4	421.9	627.2	963.3	602.8

Table 3. The X-Ray Fluorescence (XRF) component analysis of the two kinds of limestone samples.

Parameters	LOI *	CaO	MgO	SiO_2	Al_2O_3	Na_2O	Fe_2O_3	Others
Korea Samcheok	42.40	52.82	2.42	0.92	0.58	0.16	0.37	0.32
China Shengzhou	42.75	55.92	0.21	0.38	0.20	0.16	0.06	0.21

* Loss on ignition.

Before the experiment, 10–30 mg of limestone samples was uniformly dispersed on the crucible with deionized water, and then dried below 150 °C in an oven. At the beginning, the furnace was lifted to the highest height and heated to 850 °C with Ar flux through the quartz tube. After the crucible was installed and the readings of the analytical balance was stable, the heating furnace was moved downward at a speed of 5 mm/s. As shown in Figure 3, the maximum increasing rate of temperature could reach 15–20 K/s, which is much faster than that in most traditional commercial TGA, so the calcination condition in the LC-TGA is much closer to the injection condition in fluidized beds.

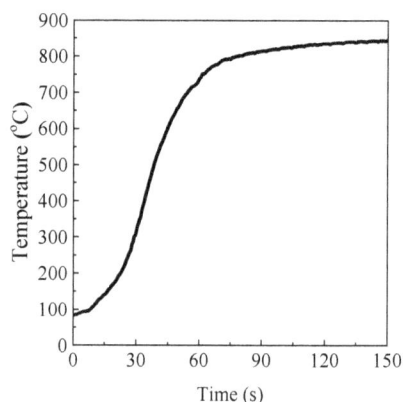

Figure 3. The heating curve of crucible in LC-TGA (5 standard liters per minute at 1 atm, 0 °C (SLM)).

Limestone samples were calcined under Ar atmosphere for 5 min and then the calcium oxide reacted with sulfur containing gas for 20 min. At the initial stage of the experiment, the mass of the limestone sample decreased quickly due to calcination. The calcination was assumed to finish after the sample mass was stable. After Ar was switched to the gas mixture of Ar, O_2 and SO_2, the sulfation reaction occurred and the sample mass increased. The O_2 concentration in the LC-TGA experiment remained unchanged at 3.5%, which is similar to that in CFB boilers. The blank experiment under the same heating rate and gas atmosphere was conducted for each set of experiments to eliminate the effects of gas flow and buoyancy on the mass measurement. With the assumption that the impurities in the sample remained constant in the reactions, the limestone conversion can be calculated by the following Equations (1)–(3):

$$n_{CaO} = \frac{m_1 \cdot LOF \cdot \gamma}{M_{CaO}} \tag{1}$$

$$n_{CaSO_4}(t) = \frac{m_3(t) - m_2}{M_{SO_3}} \tag{2}$$

$$X_s(t) = \frac{n_{CaSO_4}(t)}{n_{CaO}} \tag{3}$$

where n_{CaO} is the mole number of CaO after calcination, mol; n_{CaSO_4} is the mole number of CaSO$_4$ at a given time t, mol; M_{SO_3} is the molecular mass of SO$_3$, g/mol; m_1 and m_2 are the limestone mass before and after the calcination respectively, g; and $m_3(t)$ is the sample mass at a given time t during the desulphurization reaction, g.

The measurement results of LC-TGA was also compared with that in a fixed bed reactor (Shengfan Shiying Corporation, Lianyungang City, Jiangsu Province, China), which had the same gas controlling system. The inner diameter of the fixed bed reactor was 18 mm. Silica wool was compacted and spread on the quartz sintered distributor, preventing fine particles carried by gas flow blocking the quartz sintered distributor. Limestone samples (80 mg) were mixed well with 1.5 g quartz sands with a Saunter diameter of 150 μm, and then uniformly spread on the silica wool. During the heating process, pure CO$_2$ was introduced into the fixed bed reactor to inhibit limestone from decomposition, which was also adopted by previous studies [19]. After the furnace was heated to the given temperature (850 °C), pure CO$_2$ was switched to Ar to start limestone calcination. After 5–10 min of limestone calcination, Ar was switched to gas mixture of Ar and SO$_2$ to start CaO sulfation. The total gas flow rate was set as 2 SLM (standard liter per minute at 1 atm, 0 °C). CO$_2$ and SO$_2$ concentrations were measured by a mass spectrum analyzer with a frequency of 0.75 Hz. The sulfation conversion can be obtained by the following Equations (4) and (5):

$$n_{CaSO_4}(t) = \Delta n_{SO_2} = \frac{(P_{SO_2,0} - P_{SO_2}(t))V}{RT_A} = \frac{P_A V_A}{RT_A} \int_{t_1}^{t} (C_{0,SO_2} - C_{SO_2}(t))dt \tag{4}$$

$$X_s(t) = \frac{n_{CaSO_4}(t)}{n_{CaO}} \tag{5}$$

where P_{SO_2} is the partial pressure of SO$_2$, Pa; P_A is the atmospheric pressure, 1 atm; V_A is the total gas flow rate under the standard condition (273.15 K, 1 atm), m^3/s; R is the ideal gas constant, J/(mol·K); T_A is the atmospheric temperature, 273.15 K; $C_{SO_2}(t)$ is the outlet SO$_2$ concentration at time t, mol/mol; C_{0,SO_2} is the inlet SO$_2$ concentration, mol/mol.

In addition, a commercial TGA-Q500 produced by TA Instruments (New Castle, DE, USA) was also used to validate the measurements of the LC-TGA. The TGA-Q500 had a maximum heating rate of 50 K/min and a maximum gas volume flow rate of 200 mL/min. Limestone samples (2 mg) were used in each experiment with a gas volume flow rate of 100 mL/min. The experimental procedure for the TGA-Q500 was approximately the same as that for the LC-TGA except that the limestone was calcined at a given heating rate instead of a given environmental temperature. Two heating rates of 10 K/min and 30 K/min were both studied in the experiments.

3. Results and Discussion

3.1. Test of the LC-TGA

As shown in Figure 4, in order to eliminate the effects of external mass transfer on the sulfation reaction, experiments under different gas flow rates were compared. The sulfation conversion remained almost unchanged when the gas mixture was higher than 3 SLM. Thus, the volume flow rate of gas mixture was set to be 5 SLM in the experiments, which can eliminate the effects of external mass transfer on the sulfation reaction at different conditions. Figure 5 shows the reproducibility of the LC-TGA. It can be seen that the reproducibility is sufficient despite some data fluctuation which is lower than 0.2 mg, thus the estimated measurement data error is below ±4%.

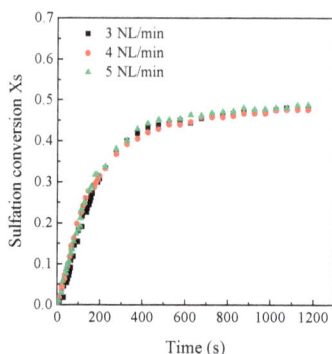

Figure 4. The sulfation conversion under different gas flow rate (Shengzhou limestone, 3.5 μm, 850 °C, 500 ppm).

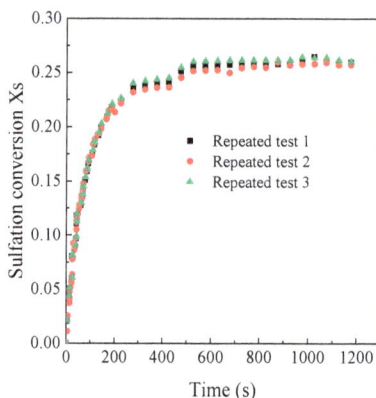

Figure 5. The repeated test results (Samcheok limestone, 26.3 μm, 850 °C, 1000 ppm).

3.2. Comparison of Sulfation Conversion in Different Reactors

The comparison of sulfation conversion in the LC-TGA and the TGA-Q500 is illustrated in Figure 6. The sulfation conversion in the LC-TGA was apparently much faster than that in TGA-Q500, although the conversion increased with the heating rate in the TGA-Q500. The time to reach a conversion of 0.5 in the TGA-Q500 was almost ten times higher than that in the LC-TGA. The high chemical reaction rate is mainly ascribed to three factors. First, limestone samples can be dispersed more uniformly in the LC-TGA with a larger crucible than that in the TGA-Q500. Particle packing can be alleviated especially for the fine powders thus the external mass transfer was improved. Second, a larger specific surface area can be obtained at a higher calcination rate in the LC-TGA, thus the porous structure promoted the sulfation reactivity. Lastly, the maximum reaction gas flow rate seemed insufficient in these cases for eliminating the external mass transfer in the TGA-Q500 due to the instrument limit, so the conversion rate was also influenced by the gas flow rate.

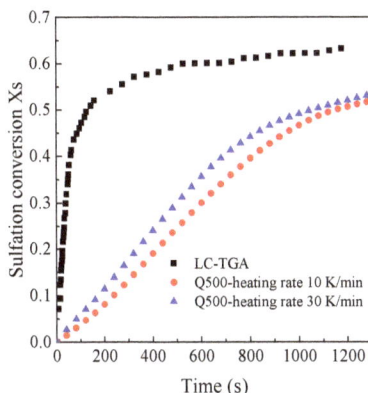

Figure 6. The comparison of sulfation conversion in the LC-TGA and the TGA-Q500 (Shengzhou limestone, 3.5 μm, 850 °C, 2000 ppm).

As shown in Figure 7, the sulfation conversions increased rather fast at the first beginning of reaction, and then turned into a slow increase both in the LC-TGA and the fixed bed reactor. However, the calcium conversion rates at the initial stage of the reactions in the LC-TGA were much faster than those in the fixed bed reactor. It is commonly known that the sulfation process is performed in two stages. The first one is fast and controlled by the chemical reaction and gas diffusion through the pore structure of particles. The second one is slower and controlled by the ion diffusion through the $CaSO_4$ production layer [22]. As the outlet SO_2 concentrations in the fixed bed reactor sharply decreased to nearly zero at the beginning of the reaction during experiments, the differential operating conditions could not be achieved. Thus, the initial CaO conversion rate in the fixed bed reactor was significantly deviated from the intrinsic reaction rate, which was limited by the external gas diffusion. In contrast, less sample mass and higher gas flow rates could be used in the LC-TGA, thus the SO_2 concentration around the CaO particles was much closer to that in the main stream, leading to a higher initial conversion rate in the LC-TGA. After the reaction goes to the second stage, the major control mechanism changes to the diffusion through the production layer, so conversion rates in both reactor systems became much closer.

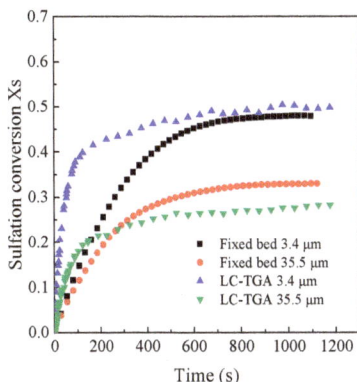

Figure 7. The comparison of sulfation conversion in the LC-TGA and the fixed bed reactor (Samcheok limestone, 850 °C, 2000 ppm).

Based on the above discussion, it can be concluded that the LC-TGA can achieve a faster calcination rate, and greatly alleviate the negative effects of the external mass transfer on the measurements of CaO sulfation reactivity, showing a good reliability for measuring the sulfation reaction with different particle sizes.

3.3. Effects of Limestone Particle Size on the CaO Sulfation Reactivity

Sulfation conversion rate with different particle sizes in the LC-TGA are shown in Figure 8. It is also observed that the final CaO conversion of the finest particle was significantly higher than that of other particle sizes. Particle size has crucial effects on the final calcium conversion. As mentioned above, the sulfation reaction blocks the surface pores, leading to an unreacted inner core. Thus, when the particle size decreased, the final CaO conversion also increased. Thus, the calcium utilization rate was improved greatly when the limestone particle size reduced to less than 20 μm or even 10 μm. In addition, the chemical reaction rates were also affected by limestone particle size. CaO conversion rates of the finer particles were higher than those of the coarser particles, especially at the very initial stage of the reactions. Thus, it can be concluded that finer limestone particles had a better CaO sulfation reactivity with higher chemical reaction rate and final conversion.

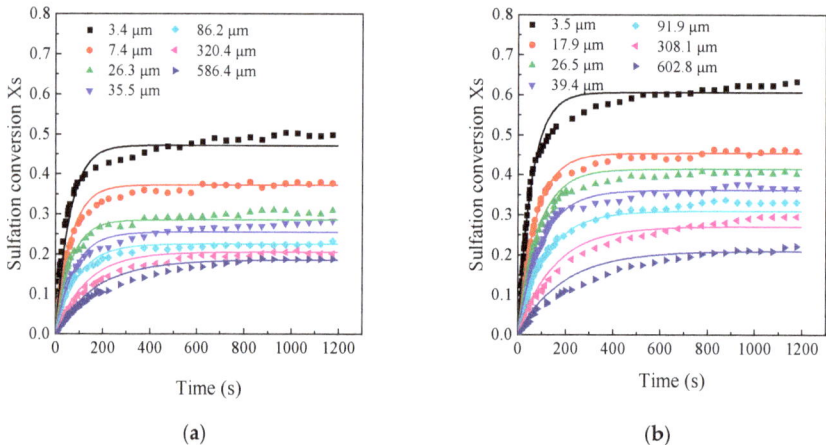

(a)　　　　　　　　　　　　　　　(b)

Figure 8. Effects of limestone particle size on the CaO sulfation reactivity (the large capacity TGA, 850 °C, 2000 ppm): (**a**) Samcheok; (**b**) Shengzhou.

3.4. Effects of SO_2 Concentration on the CaO Sulfation Reactivity

Figures 9 and 10 show the sulfation process at different SO_2 concentrations, and both the final sulfation conversion and the chemical reaction rate decreased with the SO_2 concentrations. At 250 ppm SO_2, the final CaO conversion of 600 μm particles was only 10% while the final CaO conversion of 3.4 μm particles was still as high as 40% for Samcheok limestone. The reduction of the final sulfation conversion from 2000 ppm to 250 ppm was similar for different particle sizes, thus low final sulfation conversion of the coarser limestone restricted it from realizing ultra-low SO_2 emission in CFB boilers even with the long residence time. In contrast, ultrafine limestone particles still showed a good sulfation reactivity even at low SO_2 concentrations. If the contact time can be ensured, it is more likely to realize ultra-low SO_2 emission at low Ca/S ratios by application of finer limestone.

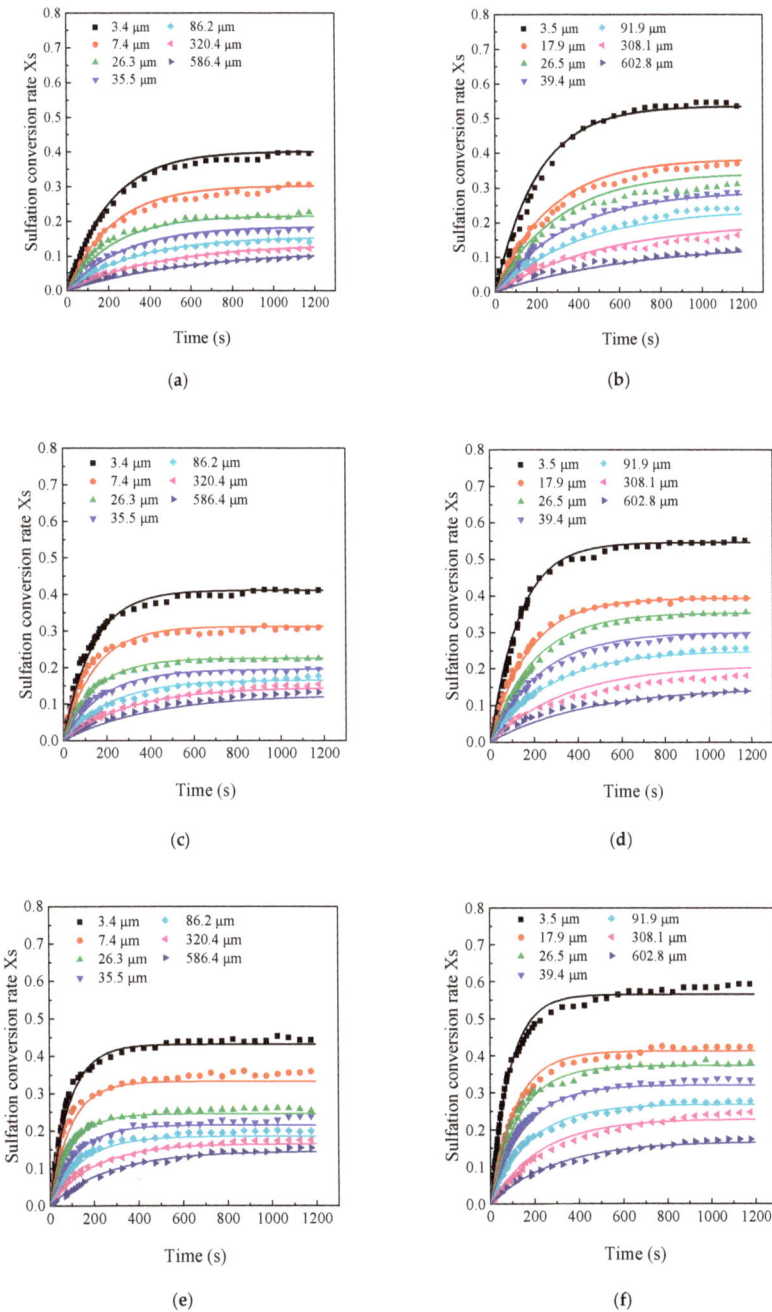

Figure 9. The sulfation conversion at different SO_2 concentrations under 850 °C: (**a**) Samcheok, 250 ppm; (**b**) Shengzhou, 250 ppm; (**c**) Samcheok, 500 ppm; (**d**) Shengzhou, 500 ppm; (**e**) Samcheok, 1000 ppm; (**f**) Shengzhou, 1000 ppm.

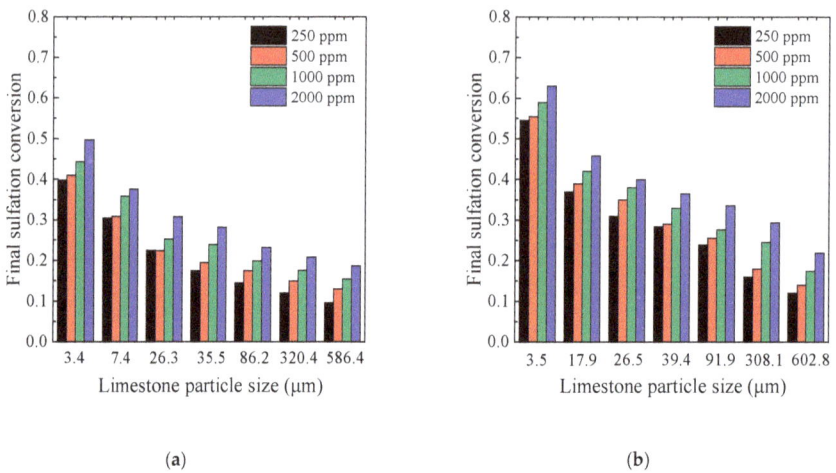

(a)   (b)

Figure 10. The final sulfation conversion at different SO_2 concentrations under 850 °C: (**a**) Samcheok; (**b**) Shengzhou.

In addition, Shengzhou limestone showed a better reactivity than Samcheok limestone, both in final conversion and reaction rate. The pore size distributions of these two kinds of limestone were measured by nitrogen adsorption apparatus ASAP 2460 produced by Micromeritics Instruments Corporation (Norcross, GA, USA), as shown in Figure 11. The measured BET (Brunauer–Emmett–Teller) specific surface areas of Shengzhou and Samcheok limestones were 38.07 m²/g and 39.23 m²/g, respectively, and the BJH (Barrett-Joyner-Halenda) adsorption cumulative pore volume were 0.177 cm³/g and 0.156 cm³/g, respectively. Although CaO particles calcined from these two kinds of limestone had similar pore surface area, the mean pore size of Shengzhou CaO was larger than that of Samcheok CaO. Previous studies have found that smaller pores will be more easily plugged and lead to the premature termination of sulfation [32]. Thus, a better pore structure of Shengzhou CaO may have enhanced its sulfation reactivity. The detailed analysis still needs further studies in the future.

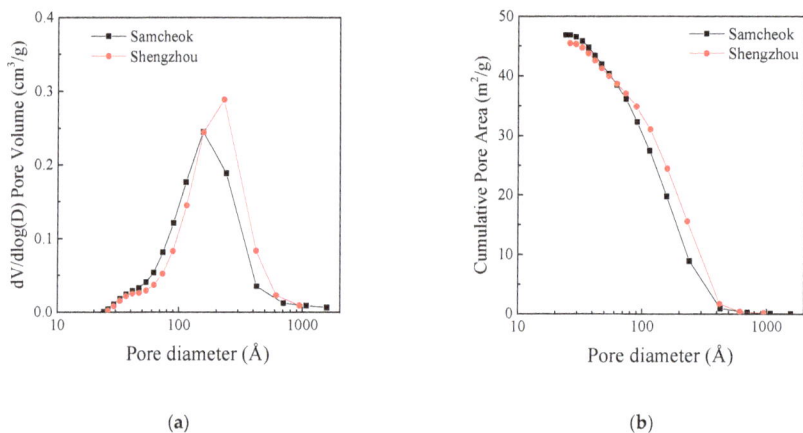

(a)   (b)

Figure 11. Pore size distribution of CaO after the calcination in the LC-TGA: (**a**) Barrett-Joyner-Halenda (BJH) adsorption dV/dlog(D) (V is the pore volume, and D is the pore size) pore volume; (**b**) BJH adsorption cumulative pore area.

3.5. Model Prediction

Many researchers have developed sulfation models to predict the conversion rate under different conditions and coupled them with gas-solid flow models to calculate desulfurization efficiencies in industrial CFB boilers. Rubiera et al. [33] proposed a classic semi-empirical model, which was widely adopted in CFB models. Two empirical parameters of $X_{s,max}$ and K_c^0 are used to predict CaO sulfation reactions as the following equation,

$$X_s = X_{s,max}(1 - \exp(-\frac{K_c^0 C_{SO_2} t}{X_{s,max}}))$$ (6)

where $X_{s,max}$ is the maximum sulfation conversion after infinite reaction time; K_c^0 is the apparent reaction rate constant at the initial reaction, m³/(mol·s); C_{SO_2} is the SO$_2$ concentration at the particle surface, kmol/m³. Thus, the conversion rate at a given time t can be calculated as the following equation,

$$\frac{dX_s}{dt} = K_c^0 C_{SO_2} \exp(-\frac{K_c^0 C_{SO_2} t}{X_{s,max}})$$ (7)

Using this model to fit with the experimental results, $X_{s,max}$ and K_c^0 of the two kinds of limestone at different SO$_2$ concentrations were obtained and listed in Table 4. $X_{s,max}$ and K_c^0 decreased significantly with the increase in particle size, which is in agreement with previous studies. Besides, it was also found that $X_{s,max}$ and K_c^0 were affected by SO$_2$ concentrations. When SO$_2$ concentrations increased from 250 ppm to 2000 ppm, K_c^0 gradually decreased and $X_{s,max}$ increased. $K_c^0/X_{s,max}$ was double at 250 ppm than that at 2000 ppm. Thus, if the model parameters obtained from the experimental results at high SO$_2$ concentrations are used to predict sulfation process at low SO$_2$ concentrations, the sulfation reaction rate may be underestimated, leading to an overestimating outlet SO$_2$ concentration. The residence time of limestone particles with a similar size as the circulating ash was sufficiently long, so the bias of this model may not be obvious at high SO$_2$ concentrations. However, as shown in Figure 12, the deviation will be much more severe at the low SO$_2$ concentration for finer limestone particles. As the residence time for the fine limestones were restricted, this model may not be satisfactory in predicting the low SO$_2$ emission at the boiler outlet.

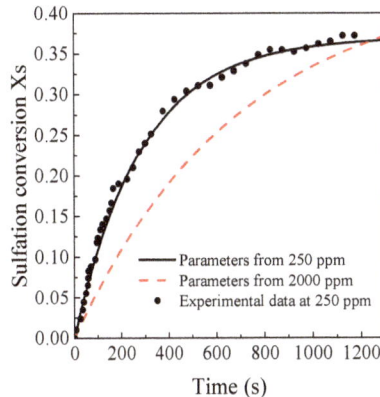

Figure 12. Predicted sulfation conversion at 250 ppm using empirical parameters from different SO$_2$ concentration using Equation (6) (Shengzhou Limestone, 17.9 μm).

Table 4. Maximum sulfation conversion after infinite reaction time ($X_{s,max}$) and the apparent reaction rate constant at the initial reaction ($K_c{}^0$) of the two kinds of limestones at different SO_2 concentrations.

		Shengzhou Limestone						
Particle size (µm)	d_{50}	7.8	37.9	65.4	93.6	165.4	362.4	627.2
	d_{32}	3.5	17.9	26.5	39.4	91.9	308.1	602.8
SO_2 concentration (ppm)								
250	$K_c{}^0$	805.4	473.4	385.7	298.0	234.0	126.9	90.3
	$X_{s,max}$	0.53	0.37	0.32	0.29	0.24	0.18	0.14
500	$K_c{}^0$	728.6	384.0	291.9	232.8	168.3	94.2	71.2
	$X_{s,max}$	0.54	0.39	0.34	0.30	0.25	0.21	0.15
1000	$K_c{}^0$	664.1	292.4	255.6	203.4	145.4	87.9	48.3
	$X_{s,max}$	0.59	0.41	0.36	0.32	0.27	0.24	0.19
2000	$K_c{}^0$	563.4	230.7	221.1	163.7	118.2	82.8	30.6
	$X_{s,max}$	0.62	0.44	0.39	0.37	0.32	0.27	0.22
		Samchoek Limestone						
Particle size (µm)	d_{50}	10.3	29.1	58.2	106.5	172.6	346.2	609.7
	d_{32}	3.4	7.4	26.3	35.5	86.2	320.4	586.4
SO_2 concentration (ppm)								
250	$K_c{}^0$	732.6	487.4	381.2	251.4	184.4	117.1	90.1
	$X_{s,max}$	0.39	0.29	0.21	0.17	0.15	0.14	0.11
500	$K_c{}^0$	673.9	415.4	311.0	213.0	145.8	100.5	58.7
	$X_{s,max}$	0.40	0.30	0.22	0.18	0.16	0.15	0.13
1000	$K_c{}^0$	560.4	327.5	290.1	205.4	130.7	87.9	41.3
	$X_{s,max}$	0.42	0.32	0.26	0.21	0.18	0.17	0.15
2000	$K_c{}^0$	437.9	262.8	177.2	166.4	108.6	66.4	36.3
	$X_{s,max}$	0.45	0.35	0.29	0.24	0.22	0.19	0.18

The modeling bias at low SO_2 concentrations means that the apparent reaction order with respect to C_{SO_2} should be lower than 1. Thus, in order to predict CaO sulfation with a broad range of particle sizes at low SO_2 concentrations, a modified empirical model was proposed in this paper as the following expression:

$$X_s = X_{s,max}(C_{SO_2}, d_p)(1 - \exp(-K_c(d_p)C_{SO_2}^m t)) \tag{8}$$

where $X_{s,max}(C_{SO_2}, d_p)$ is the final CaO conversion with a given particle diameter d_p at a given SO_2 concentration C_{SO_2}; $K_c(d_p)$ is the apparent reaction rate constant with a given particle diameter d_p, $m^{3-m}/(mol^{1-m} \cdot s)$; m is the apparent reaction order with respect to C_{SO_2}.

The apparent reaction orders with respect to C_{SO_2} of Shengzhou limestone and Korea limestone were 0.60 and 0.61 using optimal linear fitting. Borgward et al. [20] also supposed that the reaction order m with respect to C_{SO_2} should be 0.62 ± 0.07 and will be affected by limestone types. Therefore, the modification of the apparent reaction order was reasonable. As shown in Figures 13 and 14, K_c and $X_{s,max}$ for each particle size of Shengzhou limestone are plotted logarithmically against particle size d_{32}, and strong negative correlations can be obviously seen.

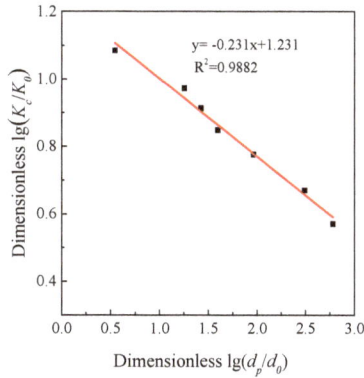

Figure 13. The relationship between apparent reaction rate constant (K_c) and particle diameter (d_p) of Shengzhou limestone.

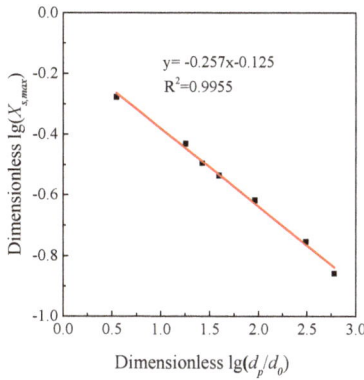

Figure 14. The relationship between $X_{s,max}$ and d_p of Shengzhou limestone (500 ppm SO_2).

The fitting reaction models of Shengzhou and Samcheok limestones are described as the following two equations, respectively.

$$X_s = \left(0.020 \times \frac{C_{so_2} - C_{so_2,0}}{C_{so_2,0}} + 0.7496 \cdot \left(\frac{d_p}{d_0} \right)^{-0.257} \right) \left(1 - exp \left(-17.036 \cdot \left(\frac{d_p}{d_0} \right)^{-0.231} \cdot C_{SO_2}^{0.61} t \right) \right) \qquad (9)$$

$$X_s = \left(0.020 \times \frac{C_{so_2} - C_{so_2,0}}{C_{so_2,0}} + 0.4834 \cdot \left(\frac{d_p}{d_0} \right)^{-0.221} \right) \left(1 - exp \left(-15.789 \cdot \left(\frac{d_p}{d_0} \right)^{-0.210} \cdot C_{SO_2}^{0.60} t \right) \right) \qquad (10)$$

where $C_{SO_2,0} = 5.4262 \times 10^{-6}$ kmol/m^3 (500 ppm); d_p is limestone particle diameter, μm; d_0 is the characteristic particle size, 1 μm.

The comparison between the experimental results and the modeling predictions are shown in Figures 8 and 9. At the initial reaction stage, the modeling predictions agreed well with the experimental results. While at the second reaction stage, there was a little bias between the modeling predictions and the experimental results. The modeling predictions showed that CaO conversion gradually approached the maximum conversion after the initial quick reaction stage. However, according to the product-layer diffusion theory, CaO still reacts with SO_2 slowly and the conversion will also increase slowly after the initial quick reaction stage. This may be the main reason for the bias of the model predictions in the later reaction stage. However, this bias seems acceptable because the increase in sulfation conversion at low SO_2 concentration was subordinate to the increase in the initial stage. Thus, it can be assumed that this empirical model can predict CaO sulfation with different particle sizes at different SO_2 concentrations, especially at low SO_2 concentrations. In addition, this model form is very simple to incorporate into available FBC models to predict SO_2 emissions for industrial applications. When using this empirical model, if it is not allowed to thoroughly study an unknown limestone in the future, it is recommended to use m = 0.6 and measure at least three characteristic particle sizes at a typical SO_2 concentration, and then the limestone reactivity can be approximately determined.

4. Conclusions

A large-capacity TGA was developed in this paper to investigate the effects of limestone particle size on the sulfation reactivity at low SO_2 concentrations, which showed a better measurement performance of the sulfation conversion than a commercial TGA-Q500 and a fixed bed reactor, especially at the initial stage of fast reaction. The experimental results showed that finer limestone particles have a better reactivity in the final conversion and faster chemical reaction rate. With the decrease of SO_2 concentration, both the final calcium conversion and the sulfation conversion rate decreased, but the ultrafine limestone particles still showed a good sulfation reactivity even at 250 ppm SO_2. If the residence time can be ensured, it is more likely for ultra-fine limestone to realize ultra-low SO_2 emissions at low Ca/S ratios.

An empirical sulfation model was established based on the experimental results. Both the final conversion and the apparent reaction rate constant had strong negative correlations against particle size, and the fitting reaction order of SO_2 was found to be about 0.6, which can well predict the sulfation process of different limestone particle sizes at low SO_2 concentrations. The model form is very simple to incorporate into available FBC models to predict SO_2 emissions.

Author Contributions: Conceptualization, Y.W., M.Z. and J.L.; formal analysis, R.C.; investigation, R.C. and Y.L.; methodology, R.C. and Y.H.; supervision, M.Z. and H.Y.; validation, R.C. and Y.L.; writing—original draft, R.C. and Y.H.; writing—review and editing, H.Z. and H.Y.

Funding: This work is financially supported by the National Natural Science Foundation of China (U1810126).

Acknowledgments: The authors thank Korea Southern Power Co., Ltd. for providing the Samcheok limestone samples.

Conflicts of Interest: The authors declare no conflict of interest.

References

1. Leckner, B. Fluidized bed combustion: Mixing and pollutant limitation. *Prog. Energy Combust. Sci.* **1998**, *24*, 31–61. [CrossRef]
2. Cai, R.; Zhang, H.; Zhang, M.; Yang, H.; Lyu, J.; Yue, G. Development and application of the design principle of fluidization state specification in CFB coal combustion. *Fuel Process. Technol.* **2018**, *174*, 41–52.
3. Yao, X.; Zhang, H.; Yang, H.; Liu, Q.; Wang, J.; Yue, G. An experimental study on the primary fragmentation and attrition of limestones in a fluidized bed. *Fuel Process. Technol.* **2010**, *91*, 1119–1124. [CrossRef]
4. Ar, I.; Balci, S. Sulfation reaction between SO_2 and limestone: Application of deactivation model. *Chem. Eng. Process.* **2002**, *41*, 179–188. [CrossRef]
5. Lyngfelt, A.; Leckner, B. Sulphur capture in circulating fluidized-bed boilers: Can the efficiency be predicted? *Chem. Eng. Sci.* **1999**, *54*, 5573–5584. [CrossRef]
6. Anthony, E.J.; Granatstein, D.L. Sulfation phenomena in fluidized bed combustion systems. *Prog. Energy Combust. Sci.* **2001**, *27*, 215–236. [CrossRef]
7. Tarelho, L.A.C.; Matos, M.A.A.; Pereira, F.J.M.A. The influence of operational parameters on SO_2 removal by limestone during fluidised bed coal combustion. *Fuel Process. Technol.* **2005**, *86*, 1385–1401. [CrossRef]
8. Braganca, S.R.; Castellan, J.L. FBC desulfurization process using coal with low sulfur content, high oxidizing conditions and metamorphic limestones. *Braz. J. Chem. Eng.* **2009**, *26*, 375–383. [CrossRef]
9. Lupiáñez, C.; Guedea, I.; Bolea, I.; Díez, L.I.; Romeo, L.M. Experimental study of SO_2 and NOx emissions in fluidized bed oxy-fuel combustion. *Fuel Process. Technol.* **2013**, *106*, 587–594. [CrossRef]

10. De Diego, L.F.; Rufas, A.; García-Labiano, F.; Abad, A.; Gayán, P.; Adánez, J. Optimum temperature for sulphur retention in fluidised beds working under oxy-fuel combustion conditions. *Fuel* **2013**, *114*, 106–113. [CrossRef]

11. Wang, C.; Chen, L. The effect of steam on simultaneous calcination and sulfation of limestone in CFBB. *Fuel* **2016**, *175*, 164–171. [CrossRef]

12. Hartman, M.; Coughlin, R.W. Reaction of sulfur dioxide with limestone and the influence of pore structure. *Ind. Eng. Chem. Process Des. Dev.* **1974**, *13*, 248–253. [CrossRef]

13. Borgwardt, R.H.; Harvey, R.D. Properties of carbonate rocks related to sulfur dioxide reactivity. *Environ. Sci. Technol.* **1972**, *6*, 350–360. [CrossRef]

14. Hartman, M.; Coughlin, R.W. Reaction of sulfur dioxide with limestone and the grain model. *AIChE J.* **1976**, *22*, 490–498. [CrossRef]

15. Fan, L.S.; Jiang, P.; Agnihotri, R.; Mahuli, S.K.; Zhang, J.; Chauk, S.; Ghosh-Dastidar, A. Dispersion and ultra-fast reaction of calcium-based sorbent powders for SO_2 and air toxics removal in coal combustion. *Chem. Eng. Sci.* **1999**, *54*, 5585–5597. [CrossRef]

16. Pisupati, S.V.; Wasco, R.S.; Morrison, J.L.; Scaroni, A.W. Sorbent behaviour in circulating fluidized bed combustors: Relevance of thermally induced fractures to particle size dependence. *Fuel* **1996**, *75*, 759–768. [CrossRef]

17. Saastamoinen, J.J. Particle-size optimization for SO_2 capture by limestone in a circulating fluidized bed. *Ind. Eng. Chem. Res.* **2007**, *46*, 7308–7316. [CrossRef]

18. Cai, R.; Ke, X.; Lyu, J.; Yang, H.; Zhang, M.; Yue, G.; Ling, W. Progress of circulating fluidized bed combustion technology in China: A review. *Clean Energy* **2017**, *1*, 36–49. [CrossRef]

19. Mattisson, T.; Lyngfelt, A. A method of evaluating limestone reactivity with SO_2 under fluidized bed combustion conditions. *Can. J. Chem. Eng.* **1998**, *76*, 762–770. [CrossRef]

20. Borgwardt, R.H.; Bruce, K.R. Effect of specific surface area on the reactivity of CaO with SO_2. *AIChE J.* **1986**, *32*, 239–246. [CrossRef]

21. Adanez, J.; Gayan, P.; García-Labiano, F. Comparison of mechanistic models for the sulfation reaction in a broad range of particle sizes of sorbents. *Ind. Eng. Chem. Res.* **1996**, *35*, 2190–2197. [CrossRef]

22. De las Obras-Loscertales, M.; de Diego, L.F.; García-Labiano, F.; Rufas, A.; Abad, A.; Gayán, P.; Adánez, J. Modeling of limestone sulfation for typical oxy-fuel fluidized bed combustion conditions. *Energy Fuels* **2013**, *27*, 2266–2274. [CrossRef]

23. Zhang, X. *Emission Standards and Control of PM2.5 from Coal-Fired Power Plant*; IEA Clean Coal Centre: London, UK, 2016.

24. Li, J.; Yang, H.; Wu, Y.; Lv, J.; Yue, G. Effects of the updated national emission regulation in China on circulating fluidized bed boilers and the solutions to meet them. *Environ. Sci. Technol.* **2013**, *47*, 6681–6687. [CrossRef]

25. Chang, S.; Zhuo, J.; Meng, S.; Qin, S.; Yao, Q. Clean coal technologies in China: Current status and future perspectives. *Engineering* **2016**, *2*, 447–459. [CrossRef]

26. Adanez, J.; Labiano, F.G.; Abánades, J.C.; De Diego, L.F. Methods for characterization of sorbents used in fluidized bed boilers. *Fuel* **1994**, *73*, 355–362. [CrossRef]

27. Simons, G.A.; Garman, A.R.; Boni, A.A. The kinetic rate of SO_2 sorption by CaO. *AIChE J.* **1987**, *33*, 211–217. [CrossRef]

28. Zarkanitis, S.; Sotirchos, S.V. Pore structure and particle size effects on limestone capacity for SO_2 removal. *AIChE J.* **1989**, *35*, 821–830. [CrossRef]

29. Milne, C.R.; Silcox, G.D.; Pershing, D.W.; Kirchgessner, D.A. Calcination and sintering models for application to high-temperature, short-time sulfation of calcium-based sorbents. *Ind. Eng. Chem. Res.* **1990**, *29*, 139–149. [CrossRef]

30. Laursen, K.; Duo, W.; Grace, J.R.; Lim, J. Sulfation and reactivation characteristics of nine limestones. *Fuel* **2000**, *79*, 153–163. [CrossRef]

31. Abanades, J.C.; Anthony, E.J.; García-Labiano, F.; Jia, L. Progress of sulfation in highly sulfated particles of lime. *Ind. Eng. Chem. Res.* **2003**, *42*, 1840–1844. [CrossRef]

Materials **2019**, *12*, 1496

32. Mahuli, S.K.; Agnihotri, R.; Chauk, S.; Ghosh-Dastidar, A.; Wei, S.H.; Fan, L.S. Pore-structure optimization of calcium carbonate for enhanced sulfation. *AIChE J.* **1997**, *43*, 2323–2335. [CrossRef]
33. Rubiera, F.; Gracia-Labiano, F.; Fuertes, A.B.; Pis, J.J. Characterization of the reactivity of limestone with SO_2 in a fluidized bed reactor. In Proceedings of the 11th International Conference on Fluidized Bed Combustion, Montreal, QC, Canada, 22–25 May 1991; pp. 1489–1495.

materials

MDPI

Article

Improved Method for Measuring the Permeability of Nanoporous Material and Its Application to Shale Matrix with Ultra-Low Permeability

Taojie Lu [1,2,3], **Ruina Xu** [1,2,3], **Bo Zhou** [1], **Yichuan Wang** [1], **Fuzhen Zhang** [1,2,3] **and Peixue Jiang** [1,2,3,*]

[1] Department of Energy and Power Engineering, Tsinghua University, Beijing 100084, China; ltj17@mails.tsinghua.edu.cn (T.L.); ruinaxu@tsinghua.edu.cn (R.X.); zhou-b12@tsinghua.org.cn (B.Z.); wyc16@mails.tsinghua.edu.cn (Y.W.); zhangfuzhen@tsinghua.edu.cn (F.Z.)

[2] Tsinghua University–University of Waterloo Joint Research Center for Micro/Nano Energy and Environment Technology, Beijing 100084, China

[3] Key Laboratory for Thermal Science and Power Engineering of Ministry of Education, Beijing 100084, China

* Correspondence: jiangpx@mail.tsinghua.edu.cn

Received: 1 April 2019; Accepted: 9 May 2019; Published: 13 May 2019

Abstract: Nanoporous materials have a wide range of applications in clean energy and environmental research. The permeability of nanoporous materials is low, which affects the fluid transport behavior inside the nanopores and thus also affects the performance of technologies based on such materials. For example, during the development of shale gas resources, the permeability of the shale matrix is normally lower than 10^{-3} mD and has an important influence on rock parameters. It is challenging to measure small pressure changes accurately under high pressure. Although the pressure decay method provides an effective means for the measurement of low permeability, most apparatuses and experiments have difficulty measuring permeability in high pressure conditions over 1.38 MPa. Here, we propose an improved experimental method for the measurement of low permeability. To overcome the challenge of measuring small changes in pressure at high pressure, a pressure difference sensor is used. By improving the constant temperature accuracy and reducing the helium leakage rate, we measure shale matrix permeabilities ranging from 0.05 to 2 nD at pore pressures of up to 8 MPa, with good repeatability and sample mass irrelevance. The results show that porosity, pore pressure, and moisture conditions influence the matrix permeability. The permeability of moist shale is lower than that of dry shale, since water blocks some of the nanopores.

Keywords: shale; permeability measurement; pressure decay method

---

1. Introduction

Porous materials are becoming increasingly popular in clean energy and environmental research, especially nanoscale porous media due to their large specific area. The permeability of nanoporous materials affects the fluid transport behavior inside the nanopores, and thus also affects the performance of technologies based on such materials [1,2].

In the last decade, shale gas has become more and more popular around the world due to its large reserves and cleanliness compared with traditional fossil fuels. The pore permeability characteristics of shale have an important influence on the determination of rock parameters, which are important to determine for the development of shale gas. The permeability represents the level to which a rock permits fluid flow at a given differential pressure. The common unit of permeability, the Darcy (D), equals 10^{-12} m². The permeability of the rock determines the ability of shale gas to flow in the pore network, which in turn, affects the development value of the corresponding reservoirs.

Unlike conventional rock systems, the pore sizes of gas-bearing shales are almost at the nanometer scale. Pores smaller than 10 nm provide most of the specific surface area and pore volume for the storage of shale gas [3,4]. Loucks et al. observed various pores with diameters of between 5 and 750 nm in the Barnett Shale [5]. Furthermore, Zhou et al. observed a large number of pores with diameters of between 5 and 200 nm in organic matter of the shale of the Longmaxi formation [6]. In addition to nanopores, shale also has micro- and nanoscale natural fractures in clay or organic matter [7].

Moreover, shale formations contain non-negligible amounts of water. On the one hand, some shale reservoirs may be damaged by the process of geological water migration. Groundwater intrusion may cause high water saturation. On the other hand, shale gas exploitation often uses hydraulic fracturing. In the process of gas reservoir reconstruction, drilling fluid, completion fluid, cementing fluid, and fracturing fluid cannot be completely discharged from the rock, meaning that a significant portion of the water injected into a well will remain in the shale rock and other geological layers [8].

Different researchers have experimentally measured the permeability of dry shale in different shale blocks, including using plunger samples and particle samples [9–13]. Wang et al. summarized the shale permeability data of nearly eight blocks in North America [14]. The shale permeability obtained from various experimental studies using different measurement methods has a wide range from 0.01 nD to 1 mD. The shale permeability measured using core plug samples is several orders of magnitude higher than that obtained using fractured particle samples, due to the existence of large natural fractures in the plugs. The natural fractures contained in artificially fractured shale particles are much smaller than those in core plug samples, so the measured permeability of such particles is more reflective of the matrix permeability.

An increase in reservoir water saturation will lead to a decrease in gas permeability and an increase in the sensitivity of the reservoir to stress damage, which is unfavorable in the later stage of shale gas extraction [15]. In laboratory research, some researchers have observed a decrease in the permeability of shale plug samples after water saturation. For example, Shen et al. found that the permeability decreases with increasing water imbibition time [16]. Additionally, Ghanizadeh et al. observed that the measured permeability coefficients of a dry shale plug were significantly higher than those measured in the same plug with 1.1% water saturation [17].

Due to the limitations of micro-flow measurement technology, the steady-state permeability test method directly based on Darcy's Law is difficult to apply in measurements of low-permeability and ultra-low-permeability materials for extremely long periods of time. For core samples with a permeability of 1 nD, the allowable air flow is only on the order of 10^{-5} cm^3/s, even if a pressure difference of 10 MPa is applied on both sides, and such micro-flow is therefore hard to measure using the traditional steady-state method. In 1968, Brace et al. presented a low-permeability core-permeability measurement technology based on the pressure decay method. Compared with the steady-state method, the pressure decay method of permeability measurement greatly shortens the test time. In the decades following 1968, Hsieh et al. [18], Neuzil et al. [19], Dicker and Smits [20], Luffel et al. [21], Jones [22], Wu et al. [23], Cui et al. [24], and Barral et al. [25] continuously developed the theory and technology of this method. In 1993, Luffel et al. proposed a pressure decay measurement principle for low-permeability shale particles [21]. By applying an initial pressure difference to a particle, unsteady one-dimensional gas flow was made to occur from the outside to the inside along the radial direction of the particle.

However, when carrying out experiments on shale with a large pore pressure, it is difficult to measure relatively small pressure changes. For example, a widely used commercial shale matrix permeability measuring instrument has a maximum allowable pressure of 1.38 MPa (200 psi), and is strongly affected by the gas tightness of helium and temperature stability. Fisher et al. sent six samples of shale to leading service companies for permeability tests [26]. The results showed that differences in the measured permeability of up to four orders of magnitude were obtained by different laboratories when analyzing the same sample. Currently, measurements of shale permeability under moist conditions are usually obtained using plug samples. While it is known that water may be easily stored in large fractures, the permeability of moist shale matrix particles is still poorly understood due to a lack of study.

This study focuses on key issues regarding the pressure decay method for the measurement of the permeability of shale particle samples. To overcome the challenge of measuring small pressure changes at high pressure, a pressure difference sensor is used in the experimental system. By improving the constant temperature accuracy and reducing the helium leakage rate, we obtain the shale matrix permeability at pore pressures of up to 8 MPa with good repeatability and sample mass irrelevance. The results show that porosity, pore pressure, and moisture condition influence the matrix permeability. The permeability of moist shale is lower than that of dry samples, since water blocks some of the nanopores.

Using this improved method, a new shale-matrix permeability measuring instrument can be designed which has a higher allowable pressure and good repeatability. During shale reservoir assessment and shale gas exploitation, this method can provide a more effective means for the measurement of permeability, which is related to the development of shale gas resources.

2. Modeling and Methods

2.1. Pressure Decay Model

The core physical process of the experiment conducted in the present study is a dynamic process of pressure decay. Porous shale particles are kept at a constant temperature, and the test gas pressure in the pores is kept stable at the initial pore pressure p_{1i}. Then, the pressure of the sample chamber is raised to p_{0i}, which results in an imbalance between the free-space pressure and pore pressure of the particles and causes a radial flow from the surface to the internal space. Finally, the pressure reaches the steady-state pressure p_f.

Unlike standard plunger samples, the size and geometry of the particles obtained by standard sieving are random variables subject to a certain statistical distribution. In order to establish a mathematical model to easily determine sample permeability, the particles are assumed to be spheres with a radius of r_0. Under this assumption, the seepage problem degenerates into an unsteady one-dimensional radial flow. Gas flows along the radial direction (r direction) of a spherical particle.

In porous media, the mass conservation equation is as follows:

$$\varepsilon \frac{\partial \rho}{\partial t} = -\frac{1}{r^2}\frac{\partial}{\partial r}(r^2 \rho u) \tag{1}$$

where ρ is the gas density and u is the Darcy velocity of radial flow, ε is the porosity of porous media. Along with Darcy's Law, the equation can be written as:

$$\frac{\partial p}{\partial t} = \frac{1}{c_g \rho \varepsilon r^2}\frac{\partial}{\partial r}\left(\frac{r^2 \rho K_e(p)}{\mu}\frac{\partial p}{\partial r}\right) \tag{2}$$

where μ is the gas viscosity, c_g is the isothermal compressibility, $K_e(p)$ is the apparent permeability, and p is the local pressure at a point inside the porous media.

The corresponding boundary conditions and initial conditions of the governing equations are:

$$\left(\frac{\partial p}{\partial t}\right)_{r=0,t} = 0 \tag{3}$$

$$p(r < r_0, t = 0) = p_{1i}, p(r = r_0, t = 0) = p_{0i} \tag{4}$$

$$V_f\left(\frac{\partial p}{\partial t}\right)_{r=r_0,t} = 4\pi r_0^2\left(-\frac{\rho K_e(p)}{\mu}\frac{\partial p}{\partial x}\right)_{r=r_0,t} \tag{5}$$

where V_f is the volume of free space. Equation (5) describes how the gas in free space enters into the particle pore in the process of gas permeation.

Several researchers have provided solutions to such pressure decay equations. For instance, Profice et al. combined mass conservation with the Klinkenberg equations to give the solution [27]. Cui et al. developed an analytical method to calculate shale permeability [24]. They suggested that the slope of the formal stage of the logarithmic curve can be used to determine the permeability. Within the range of experimental pressure, the gas viscosity μ can be approximated as a constant. The test gas is assumed to be an ideal gas. Additionally, we define a non-dimensional number f which reflects the ratio of pore volume to cavity free-space volume:

$$f = \frac{\varepsilon r_0 (4\pi r_0^2)}{V_f} = \frac{3V_p}{V_f} \tag{6}$$

where V_p is the pore volume. Each volume satisfies the pressure equilibrium relation:

$$p_{0i}V_f + p_{1i}V_p = p_f(V_f + V_p) \tag{7}$$

The apparent permeability K_e (p) does not change during a single experiment as the pressure decay [24]. The dimensionless analytic solution of Equation (2) is:

$$p(\tilde{r}, \tilde{t}) - p_f = (p_{0i} - p_{1i}) \sum_{n=1}^{\infty} \frac{2\sin(\varphi_n \tilde{r}) \exp(-\varphi_n^2 \tilde{t})}{\tilde{r}[\varphi_n \cos \varphi_n + (2+f) \sin \varphi_n]} \tag{8}$$

where dimensionless radius and time are respectively:

$$\tilde{r} = r/r_0 \tag{9}$$

$$\tilde{t} = t/t_0, t_0 = \frac{r_0^2 \varepsilon \mu c_g}{K_e(\bar{p})}, \bar{p} = \frac{1}{2}(p_{0i} + p_{1i}) \tag{10}$$

and ϕ_n are the characteristic roots of the following Equation (11). In one experiment, we define an average pressure to measure the permeability under this pressure, which is the mean value of the initial pore pressure p_{1i} and the sample chamber pressure p_{0i}. As f decreases, each characteristic root tends to be a positive integer multiple of π, as shown in Figure 1.

$$\tan \varphi = \frac{f\varphi}{\varphi^2 + f} \tag{11}$$

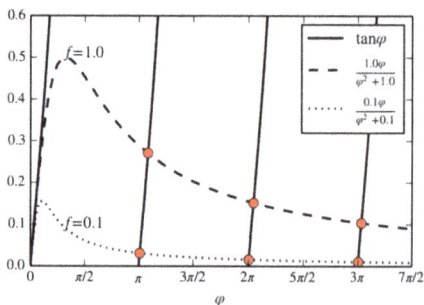

Figure 1. Graphical representation of the characteristic Equation (11). The solid lines represent the tangent function of ϕ and the dotted lines represent the fraction with different values of the non-dimensional number f. The red circles represent the characteristic roots of Equation (11). As f decreases, each characteristic root tends to be a positive integer multiple of π.

When the dimensionless time is greater than 0.1, the result calculated with the first characteristic root is almost consistent with the result obtained by taking the first 100 characteristic roots, as shown in Figure 2, which indicates that the first characteristic root dominates the whole pressure decay process. By solving the dimensionless equation of the pressure decay model, when the pressure decay time is over $0.1t_0$, we can use the slope of the logarithmic curve to calculate the permeability.

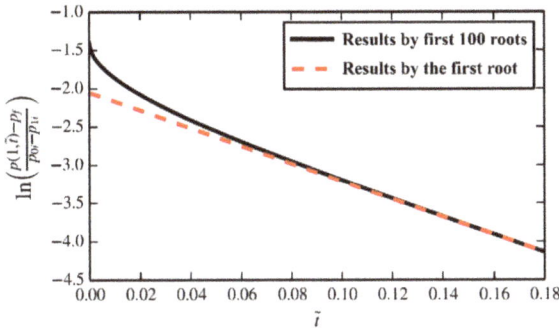

Figure 2. The truncation result of the analytic solution. The solid line represents the result of the truncation for the first 100 characteristic roots of Equation (11) and the dotted line represents the result of the truncation for the first characteristic root. When the dimensionless time is greater than 0.1, there is little difference between the two curves.

When only the first characteristic root is calculated, the function degenerates into an exponential function, and the fitting slope S of logarithmic pressure with time can be obtained:

$$|S| = \frac{\varphi_1{}^2}{t_0} \tag{12}$$

$$K_e = \frac{|S|r_0{}^2 \varepsilon \mu c_g}{\varphi_1^2} = \frac{|S|r_0{}^2 \varepsilon \mu}{\varphi_1^2} \frac{1}{\bar{p}} \tag{13}$$

2.2. Experimental Method

The experimental system for the measurement of shale matrix particle permeability is shown in Figure 3. The experimental system consists of a high-pressure helium source, a reference chamber, a sample chamber, a voltage regulator, a thermostatic water bath (accuracy ±0.05 °C), a pressure sensor, and other essential components which are connected by high-pressure sealing pipes. The helium cylinder is connected to an inlet pressure sensor (model EJA430A, Yokogawa Electric Corporation, Tokyo, Japan; range 0.14–14 MPa, accuracy 0.065%) via a pressure-reducing valve. An electric control valve is arranged on the pipeline connecting the reference chamber and the sample chamber. A high-pressure differential pressure transducer (model EJA110A, Yokogawa Electric Corporation; range 0.5–10 kPa, accuracy 0.075%) is set in parallel with the electric balance valve to record the pressure difference between the reference chamber and the sample chamber. A platinum resistance temperature sensor (accuracy ±0.1 °C) is placed in the reference chamber to record the temperature in the system. The entire core line is placed in the thermostatic water bath. To avoid current fluctuation, a voltage stabilizer is set along with the water bath.

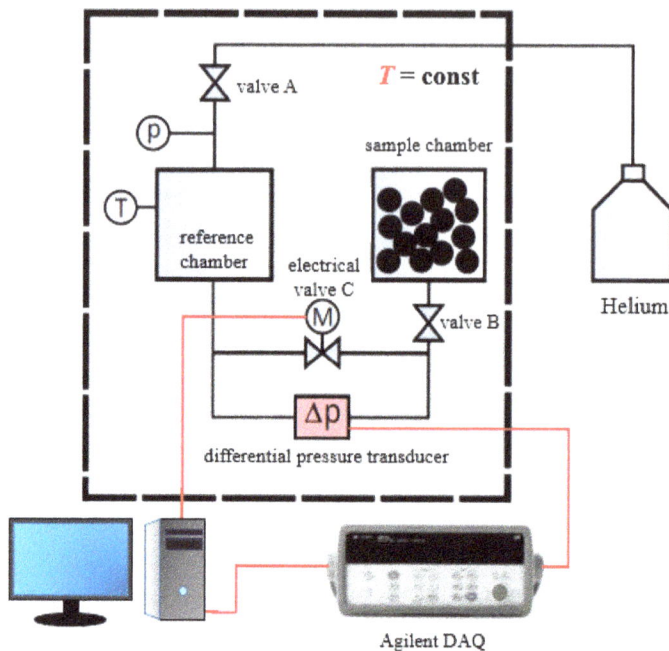

Figure 3. Schematic diagram of the experimental system used for the measurement of shale matrix particle permeability. Shale particle samples are placed in the sample chamber, and a reference chamber with approximately the same volume as the sample chamber is used for gas buffering. The electrical valve C is automatically controlled by the computer. The whole system is placed in a thermostatic water bath to maintain a constant temperature.

Before the experiment, shale samples were crushed and passed through a standard sieve. A certain mass of particle samples was dried for over 12 h and then put into the sample chamber. The chamber volume is almost equivalent to that of the reference chamber. After securing the helium tightness of the device, the whole system was placed in the thermostatic water bath, and helium was pumped into the reference chamber at a certain pressure. After the system pressure had stabilized, we opened the sample chamber valve, causing high-pressure gas in the reference chamber to enter the low-pressure sample chamber. A pressure pulse is generated due to the valve action. When the whole free-space pressure is balanced at p_{0i}, the electric control valve will automatically identify the pressure pulse and close the valve after a delay time t_w. When the valve closes, the two chambers are separated; the pressure of the reference chamber is stable at p_{0i}, while that of the sample chamber decays from p_{0i} to p_f. The differential pressure sensor records the pressure difference between the two chambers. The diagram of pressure change is shown in Figure 4.

After the sample chamber valve is opened, the whole system will go through three main stages:

I. Free-space pressure balance After the two chambers are connected, gas flows into the sample chamber from the reference chamber and the pipeline through the electric balance valve. When the pressure in the whole free space is basically balanced to p_{0i}, the signal from the pressure difference transducer returns to zero. The duration of the entire free-space pressure balancing process depends on the pipe volume and flow resistance of the system.

II. Free-space gas infiltration in both chambers As the gas gradually begins to infiltrate into the particle sample, the pressure of the free space drops. Until the electric balance valve is cut off, the output of the pressure difference sensor is always zero.

III.   Sample chamber free-space gas infiltration After the electric balance valve is cut off, only the gas in the sample chamber can infiltrate into the sample. The differential pressure curve of the two chambers, $\Delta p(t)$, is recorded by the differential pressure sensor. The differential pressure curve increases until the end of the pressure decay process, and the final output of the differential pressure sensor is denoted as Δp_f.

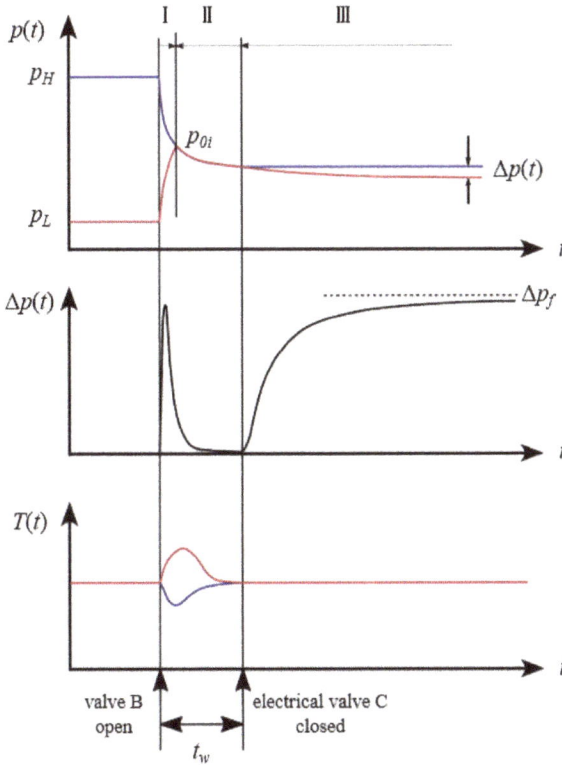

Figure 4. Graphs of the pressure $p(t)$ of the reference chamber and sample chamber, the pressure difference signal $\Delta p(t)$ measured by the differential pressure transducer, and the gas temperature $T(t)$ in the pipe.

2.3. Experimental Materials

In this study, the aforementioned experimental system was used to measure the matrix permeability of four groups of shale samples from the Longmaxi formation.

Before the experiment, the petrophysical properties related to porosity and permeability were measured by the PetroChina Research Institute of Petroleum Exploration and Development in Langfang. The results are summarized in Tables 1 and 2. TOC means the Total Organic Carbon. More data and figures can be seen in Appendix A Figures A3–A6.

Table 1. Petrophysical properties of the shale particle samples.

Serial Number	Depth (m)	TOC (%)	Sulfur (%)	Porosity (%)
1	2321.66	2.1	2.5	2.14
2	2329.76	3.9	4.8	2.90
3	2338.84	4.1	6.3	3.62
4	2346.01	5.9	4.2	4.54

Table 2. Specific surface area and pore size distribution of the shale particle samples.

Serial Number	Specific Surface Area [1] (m^2/g)	Cumulative Pore Volume [2] (cm^3/g)	Average Pore Diameter A [1] (nm)	Average Pore Diameter B [2] (nm)
1	18.4309	0.028707	6.55	9.32
2	23.9007	0.035465	6.33	9.26
3	23.4238	0.028691	5.47	8.28
4	24.6843	0.026502	4.80	7.25

[1] Data processed with the BET method. [2] Data processed with the BJH method.

2.4. Data Processing

The sample particles were passed through sieves with different mesh sizes. Each experiment was kept under average pressure for more than 2000 s. The constant temperature was set at 30 °C. For the experiment, 20–35 mesh shale particles with a mass of 10 g were chosen. The initial pressure was maintained at around 1 MPa and the average pressure changes were kept within the range of 1–8 MPa.

By calculating the pressure decay curve logarithmically, the pressure decay process goes into the formal stage after the initial time. Using the result of the pressure decay model equation, the permeability under different pressures can be calculated using the slope fitted for the linear stage of the logarithmic curve. An example of the data processing of sample 1 is described in Appendix A.

3. Results

3.1. Repeatability of Experiment and Mass Influence

Firstly, we used sample 1 to determine the repeatability of the experiment. For the experiment, 20–35 mesh shale particles with a mass of 10 g were chosen. The initial pressure difference was maintained at around 1 MPa and the average pressure changes were kept within the range of 1–8 MPa. Figure 5 shows the permeability values obtained from five trials using sample 1. The difference in the matrix permeability determined in the five experiments was smaller than 1 nD. When the average pore pressure increases, this difference gradually reduces; at a pore pressure of 8 MPa, the difference is almost within 0.1 nD. The results of the repeatability experiments demonstrate the stability of the permeability measurements obtained using the experimental system.

Secondly, we investigated the effect of sample quality on the shale matrix permeability. For the four groups of samples, 6, 10, and 14 g of 20–35 mesh particle samples were respectively put in the sample chamber, and experiments were carried out under a pressure of 3 MPa and a temperature of 30 °C. Figure 6 shows the pressure difference signal and logarithmic pressure decay curve for sample 1.

Figure 7 shows the permeability values of the four groups of samples for different sample masses. For the same sample group, the permeability differs by less than 0.2 nD for different sample masses under the same experimental conditions. The results show that the matrix particle permeability is not sensitive to the change of sample mass.

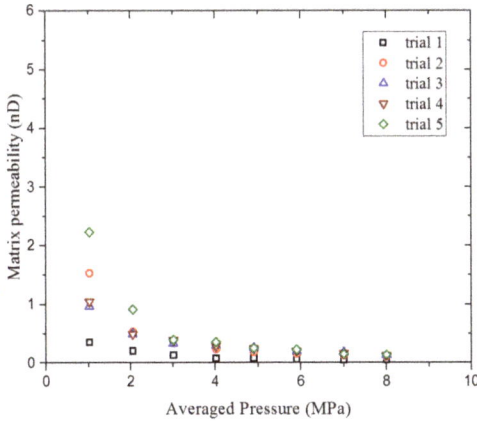

Figure 5. The results of the repeatability experiment, involving five trials, using sample 1 under pressures of 1–8 MPa.

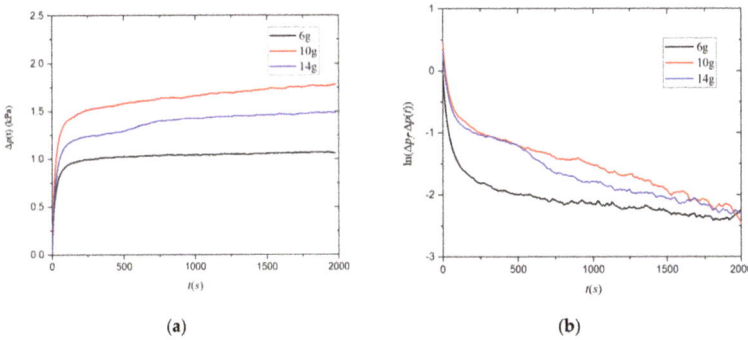

(a)   (b)

Figure 6. (**a**) The pressure difference signal $\Delta p(t)$ and (**b**) logarithmic pressure decay curve for sample 1 measured at a pressure of 3 MPa and a temperature of 30 °C The black, red, and blue lines indicate sample masses of 6, 10, and 14 g, respectively.

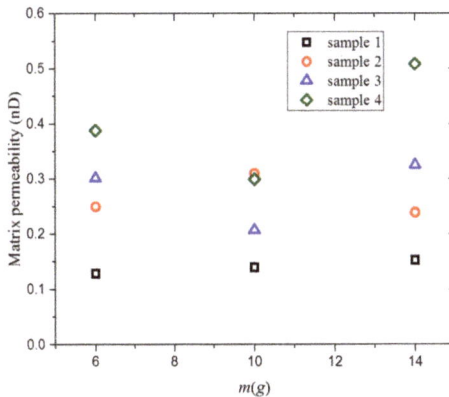

Figure 7. The permeability values of four groups of shale samples for sample masses of 6, 10, and 14 g measured under a pressure of 3 MPa and a temperature of 30 °C.

3.2. The Effects of Pressure on Permeability

To investigate the effect of pressure on the estimated permeability, we chose a total of 20–35 mesh shale particles with a mass of 10 g. The initial pressure difference was maintained at around 1 MPa and the average pressure changes were kept within the range of 1–8 MPa. Figure 8 shows the permeability values of the four groups of samples under different pressure conditions. It can be seen from the figure that the overall permeability of the shale matrix particles decreases with increasing pressure, especially at low pressure; this corresponds to the Klinkenberg effect.

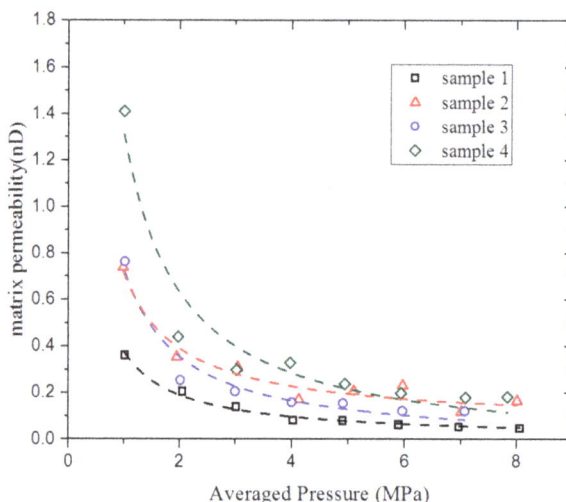

Figure 8. Matrix permeability of four shale particle samples under different pressures. The dotted line is the result of the fitting by the Klinkenberg equation.

The fundamental reason why the permeability of shale matrix is different from that of conventional reservoir rocks is the nanoscale characteristics of fluid flow. Under the temperature and pressure of shale reservoirs, the flow of methane gas in the nanopores is a rarefied gas flow, as shown in Figure 9. The Knudsen number describes the ratio of the mean free path of a gas molecule λ and the characteristic flow length l_f:

$$K_n = \frac{\lambda}{l_f} \tag{14}$$

In the nanoscale pores of shale, the flow state of gas is mainly slip flow and transition flow. In the slip zone, the gas retains the characteristics of the continuous medium, however has a non-zero velocity on the solid boundary.

In 1945, Klinkenberg studied the slip flow of low-permeability rock caused by the gas rarefaction effect, and found that the rock permeability measured by liquid medium was a constant independent of pressure, while the permeability measured by gaseous medium increased with decreasing experimental pressure. Klinkenberg concluded that the slip velocity of gas on the rock pore surface caused the correlation between permeability and pressure. He derived the following Klinkenberg equation based on a laminar flow model considering slip velocity in a one-dimensional uniform pipe [28]:

$$K_e = K_\infty \left(1 + \frac{b_k}{p}\right) \tag{15}$$

where K_e is the apparent permeability of the rock, K_∞ is the absolute permeability of the rock, and b_k is the Klinkenberg slip constant. Different researchers have carried out experimental fitting and theoretical derivation for the slip constant in Equation (15) [9,29–31].

The fluid property in confined nanopores is different from the large-scale fluid property, due to the influence of capillarity, interfacial phenomena, and hydrodynamics. For regularly shaped structures such as hard-sphere nanopores or nanochannels, theoretical and simulation methods, such as density functional theory and Monte Carlo simulation, are effective ways to describe the fluid mechanics [32–34].

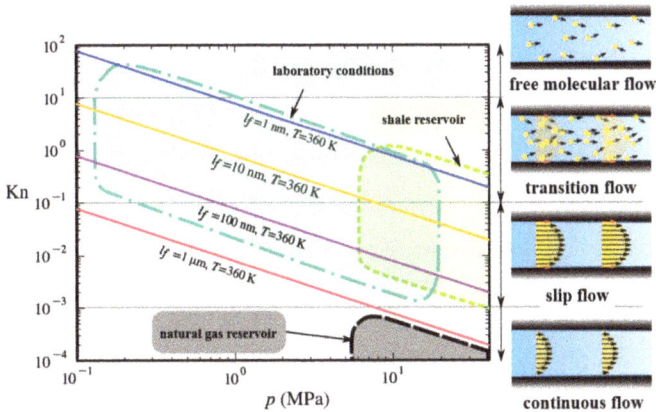

Figure 9. Rarefied gas flow type under different conditions. This can be divided into free molecular flow ($K_n > 10$), transition flow ($0.1 < K_n < 10$), slip flow ($10^{-3} < K_n < 0.1$), and continuous flow ($K_n < 10^{-3}$).

In our study, since the pressure ranges from 1 to 8 MPa, the flowing gas is in the transition zone and slip zone. The gas maintains the properties of a continuous medium with a velocity at the boundary of the solid, so the basic equations are still used. Further research is warranted into the apparent permeability of shale matrix, the absolute permeability model, and the more acute fluid property in confined nanopores.

3.3. Permeability of Moist Shale Particles

The moist permeability was measured for shale particle sample 1 used in the previous dry sample experiment. After a long period of vacuum treatment, we added water to the vessel and left it for over 48 h to allow saturation. Water on the surface of the rock was then removed by filter paper. By the weighing method, the water saturation rate was determined to be 1.5%.

To analyze the differences between the two samples, the particles between the dry and moist samples were scanned with a Nuclear Magnetic Resonance (NMR) instrument. Low-field NMR is a non-destructive technique for the quantitative measurement of hydrogen-containing fluids in porous media. Here, we use the T_2 time to reflect different transverse relaxation times of hydrogen-containing media of different scales in pores. A shorter T_2 time corresponds to a smaller amount of fluid. Researchers have demonstrated that a T_2 of around 1 ms contains a signal from nanopores in kerogen [35–37].

The change in the transverse relaxation amplitude after water saturation is shown in Figure 10. For dry samples, we consider that after a long drying time, the remaining shale gas in the pores had almost all escaped, and the signal comes from the solid skeleton of a rock sample. In NMR signal the peak of relaxation amplitude appears between T_2 values of 0.1–1 ms, which we believe represents the water molecule in the nanopores of the shale rock.

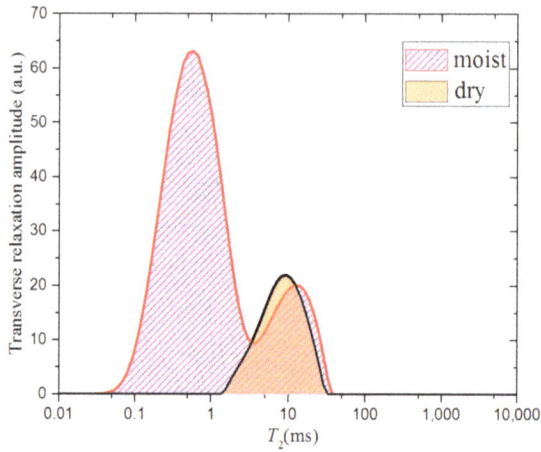

Figure 10. Transverse relaxation amplitude of shale particle sample 1 detected by low-field Nuclear Magnetic Resonance. The T_2 measuring sequence is CPMG.

The permeability of moist samples were measured in a similar way to dry samples. The permeability of the same sample decreases after water saturation under the same pressure, as shown in Figure 11. The two permeability vs. pressure curves (i.e., for dry and moist samples) were fitted using the Klinkenberg equation. It can be seen from the fitting relations that the absolute permeability of the dry and moist samples differs by about 0.04 nD. According to the fitting relation, with increasing pressure, the influence of the moisture content on the matrix permeability gradually increases, with a maximum decrease in permeability of 30% being observed.

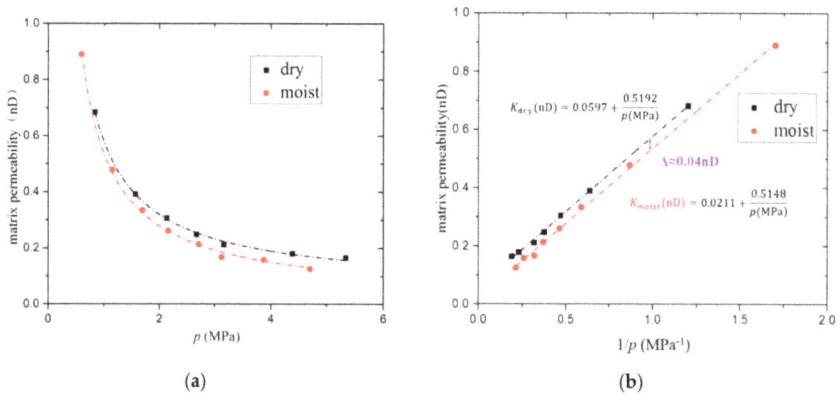

(a)

(b)

Figure 11. Matrix permeability of dry and moist (1.5% water) samples of shale sample 1 under different pressures. (**a**) Matrix permeability vs. pressure. The dotted lines are the results of fitting using the Klinkenberg equation. (**b**) Matrix permeability vs. $1/p$.

4. Discussion

When using the pressure decay method to measure the permeability of crushed samples, many researchers have extensively used commercial equipment. Achang et al. found that temperature regulation was the key to obtaining consistent results [38]. If the temperature fluctuates, the available data from the pressure decay curve is limited. This requires the experimental system to work steadily under constant temperature for a certain period of time. An air thermostat is usually used to maintain the temperature of the sample chamber. Air thermostats have the advantages of convenience and the fact that they have less impact on the design of the system. However, due to the smaller specific heat of air compared with water, the constant temperature measurement of an air thermostat may be slightly more unstable than that obtained using a similarly priced water bath. In our experiment, the temperature was maintained to within ±0.05 K.

Although the pressure decay method provides an effective way to measure low permeability, commercial apparatuses have difficulty measuring low permeability at high pressure conditions over 1.38 MPa. Therefore, Klinkenberg corrections and in situ effective stress corrections are essential to relate the results to real reservoirs [27,39–41]. Some laboratories run pressure decay tests at different pressures. However, the measured pressure curve consists of a number of discrete steps (e.g., Heller et al. [42]), which leads to difficulties in data processing. This is due to the fact that, when using a system based on the Luffel system, in which one pressure meter is used to record the whole system pressure, the total variation of the pressure decay curve is very small at a high pressure range, and is almost equal to the resolution of the pressure measurement. If one tries to use a system based on the Luffel system to measure the permeability of low-permeability rocks under a pressure of 10 MPa, the pressure change in the sample chamber will be quite small, and so the measured pressure curve will consist of discrete steps. Therefore, in this study, we attempted to use a pressure difference sensor that resists high pressure to measure the pressure change in the sample chamber. Since the pressure range of the differential pressure sensor is almost the same order of magnitude as the variation of the pressure decay curve, the measurement accuracy of the experimental system will be improved, and the measurement accuracy does not change with increasing system pressure within the pressure limit.

In pressure decay experiments on low-permeability particles, helium is often used as a test gas to reduce the effect of adsorption. The molecular weight of helium is very low, and tight sealing conditions are therefore required. Some researchers conducted a leak test for a commercial permeability-measuring apparatus, and recorded a leakage rate of <0.062 psi/s [24]. In our experimental system, a better sealing system is designed for the sample chamber; even at a pressure of 8 MPa, the leakage rate is lower than 0.025 kPa/min, which ensures the accuracy of the pressure measurement in the experiment.

In our study, matrix permeability showed little reduction after water saturation. However, natural shale formations have a multi-scale pore structure. In such formations, large cracks tend to be blocked easily, which is the main factor that leads to the reduction of permeability. Especially in low-permeability and ultra-low-permeability reservoirs, the pore throats, which allow the free flow of fluid, are small, and the capillary pressure is therefore large, which makes it more difficult to discharge water. Additionally, the spherical bubbles in oil or oil droplets in water need to overcome additional resistance due to interfacial tension during oil displacement. The water blocking effect and Jamin effect are disadvantageous to the exploitation of oil and gas resources. The damage index varies with the conditions of the formation. To better analyze the impact of moisture on permeability, a multi-parameter porosity and permeability model is needed [43,44].

In this study, permeability information for the shale matrix samples was calculated by analyzing the experimentally derived pressure decay curve. In the data processing, errors come from the accuracy of the model and the measurement of parameters.

When ignoring the effect of adsorption in the mathematical model, physical effects such as gas compressibility will affect the calculation of the apparent permeability. Here, we use a numerical method based on a finite-difference scheme to calculate the relative errors at different permeabilities due to ignoring compressibility, which are shown in Figure 12. With increasing initial pressure ratio $\bar{p}/p_{1i}$, the error due to ignoring compressibility increases. On the other hand, increasing the value of f can reduce the error under certain pressure conditions. This is due to the fact that, when the value of f is smaller, the first characteristic value ϕ_1 is closer to π. Therefore, the pressure inside the particle will approach the final pressure p_f more quickly, and p_f will be close to p_{0i}. In summary, for systems with a smaller dimensionless number f, when the system initial pressure ratio $\bar{p}/p_{1i}$ is large, ignoring the gas compressibility will lead to a larger error in the calculated permeability.

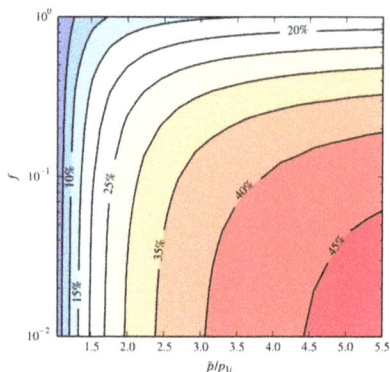

Figure 12. The relative error in the calculated permeability due to ignoring gas compressibility. For systems with a smaller dimensionless number *f*, when the system initial pressure ratio is large, ignoring the gas compressibility leads to a larger error in the calculated permeability.

Additionally, the apparent permeability in low-permeability porous media is related to pressure. Assuming that the apparent permeability $K_e(p)$ does not change in a single experiment as the pressure decays leads to an error from ignoring the rarefaction effect. Considering the dimensionless equation:

$$\frac{\partial p}{\partial \tilde{t}} = -\frac{1}{\mu} \frac{1}{\tilde{r}^2} \frac{\partial}{\partial \tilde{r}} (\tilde{r}^2 \frac{K_e(p)}{K_e(\overline{p})} \frac{p}{\overline{p}} \frac{\partial p}{\partial \tilde{r}}) \tag{16}$$

where the permeability radio $K_e(p)/K_e(\overline{p})$ has the same status with pressure ratio $p/\overline{p}$ in the approximate solution. Thus, similar to the error analysis for gas compressibility, for systems with smaller dimensionless number *f*, when the permeability ratio $K_e(\overline{p})/K_e(p_{1i})$ is large, ignoring the rarefaction effect of gas will lead to a larger error in permeability.

In our study, we mainly use shale particles of around 20 mesh, the diameter of which is around 1 mm. Larger particles may not meet the assumption of sphericity, so that the one-dimensional model cannot be applied. When the particle size continues to increase, there may be a small number of macropores in the particle, which changes the permeability of the particles. Cui et al. [24] used a dual-porosity model which takes into account the changes in particle permeability caused by micropores and macropores, and obtained an empirical formula for the permeability. The equivalent permeability (K_e) determined using the methods proposed in the present study should be related to the permeability of macropores (K_a) and micropores (K_i) as follows:

$$\frac{1}{K_e} = \frac{1}{K_a} + \frac{C_0 R_i^2}{R_a^2 K_i} \tag{17}$$

where C_0 is a constant related to the size of the micropores in the particles (R_i), rock particle size (R_a), macroporosity, and microporosity.

The Klinkenberg model describes a theoretical permeability model as a function of the gas pressure. However, the Klinkenberg relation was derived based on the formula of the mass flow rate in cylindrical tubes [28]. It can be seen from the formula that the permeability will increase to infinity at low pressure, which is not physically possible. The reason for this failure is the assumption that the porous medium is a bundle of tubes with uniform cross-sections. Zhou et al. improved the Klinkenberg permeability model for porous media with pore-scale wall-slip considering pore geometry complexity [45], creating an effective permeability model, called the general slip regime (GSR) model. The model can be expressed as follows:

$$K_e = K_\infty \left(\frac{1 + S_1 \sigma K_n}{1 + S_2 \sigma K_n} \right) \tag{18}$$

where S_1 and S_2 are REV geometry dependent rank 2 tensors, and K_∞ is the liquid permeability which does not involve the wall-slip effect for gas flow. When Kn < 0.01, the GSR model can be linearized to the Klinkenberg model. When Kn ≤ 0.1, the Klinkenberg model of apparent permeability is overvalued by 15–70%. Here, we use the GSR model to fit the permeability results by pressure for comparison with the Klinkenberg model, as shown in Figure 13. At low pressure, the permeability obtained using the GSR model is lower than that obtained using the Klinkenberg model. When the characteristic flow length is small, the Knudsen number is larger and the difference in the permeabilities obtained by the two models is more pronounced.

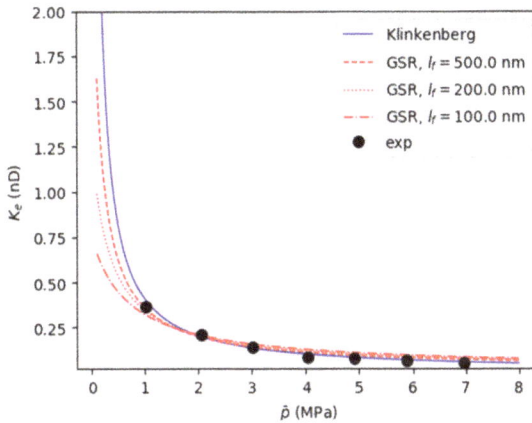

Figure 13. A comparison of the permeability fitting using the general slip regime (GSR) model and the Klinkenberg model. The characteristic flow scales l_f used for the fitting with the GSR model were 100, 200, and 500 nm.

5. Conclusions

An experimental investigation of the permeability of shale matrix was conducted using the pressure decay method. The conclusions of this investigation are as follows:

1. The proposed experimental method for the measurement of low permeability represents an improvement over previous methods. To overcome the challenge of measuring small pressure change at high pressure, a pressure difference sensor is used. By improving the constant temperature accuracy and reducing the leakage rate of helium, we obtain the shale matrix permeability at pressures of up to 8 MPa and pore pressures ranging from 0.05 to 2 nD, with good repeatability and sample mass irrelevance.
2. As gas molecules inside nanopores are affected by the Klinkenberg slip effect, the apparent permeability is larger when measured at low pressure. With increasing pressure, the permeability measured under high pressure is closer to the absolute permeability of the particles.
3. The permeability of moist shale is lower than that of dry shale, since water blocks some of the nanopores. In natural shale formations, large cracks tend to fill with water more easily, which leads to the reduction of permeability.

Author Contributions: Conceptualization, P.J. and R.X.; Funding acquisition, R.X., P.J. and F.Z.; Investigation, T.L., B.Z., Y.W., and F.Z.; Methodology, T.L., B.Z., and Y.W.; Writing—original draft, T.L.; Writing—review and editing, R.X. and P.J.

Funding: This research was funded by Major National Science and Technology Projects (2017 ZX05035-002). The authors acknowledge support from the National Natural Science Foundation of China for Excellent Young Scientist (Grant No. 51722602). This research is also sponsored by the Creative Seed Fund of Shanxi Research Institute for Clean Energy, Tsinghua University.

Acknowledgments: We thank the PetroChina Research Institute of Petroleum Exploration and Development in Langfang for their help in the characterization of petrophysical properties and pore size distribution.

Conflicts of Interest: The authors declare no conflict of interest.

Appendix A

Here, we take sample 1 as an example to describe the data processing procedure. Figure A1 shows the pressure difference curve detected by the pressure difference sensor under different equilibrium pressures.

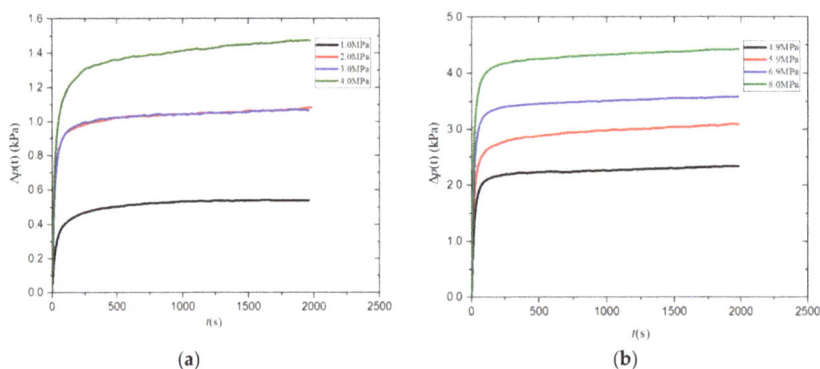

Figure A1. Pressure difference signal ($\Delta p(t)$) for shale particle sample 1 measured by the differential pressure transducer under pressures of 1–8 MPa. (**a**) average pore pressure under 1–4 MPa with smaller Δp_f; (**b**) average pore pressure under 5–8 MPa with larger Δp_f.

After the pressure pulse, the pressure inside the sample chamber decreases continuously, and the pressure difference between the sample chamber and the reference chamber gradually increases. By displaying the pressure decay curve logarithmically, as shown in Figure A2, it can be seen that at the early stage of the pressure decay process, the free-space gas rapidly enters the pores of the matrix particles, and the curve drops sharply. Then, the gas infiltration and migration slows down, and the curve tends to be linear. According to the dimensionless result, when the dimensionless time is greater than 0.1, the pressure decay process goes into the formal stage. Using Equation (13), we can calculate the sample permeability under different pressures using the slope fitted at the linear stage of the logarithmic curve.

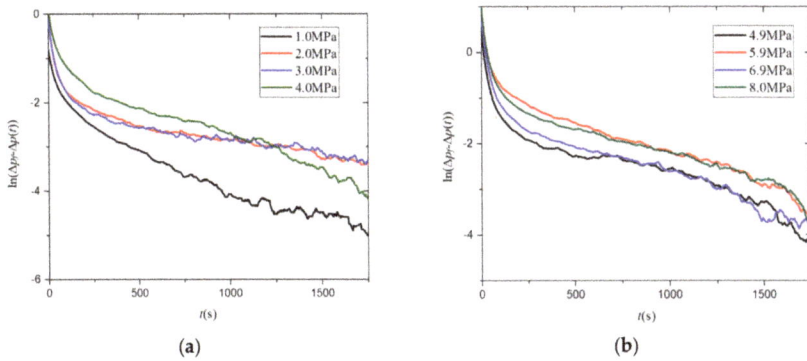

Figure A2. Logarithmic pressure decay curve, $\ln(\Delta p_f\text{-}\Delta p(t))$, for shale particle sample 1 under pressures of 1–8 MPa. (**a**) average pore pressure under 1–4 MPa; (**b**) average pore pressure under 5–8 MPa.

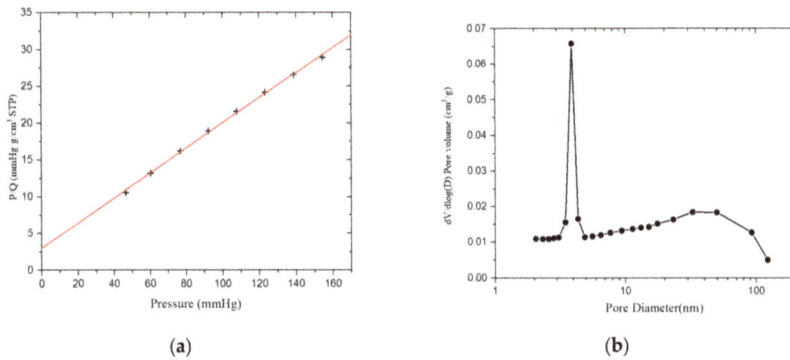

Figure A3. Nitrogen adsorption of sample 1 measured by an ASAP 2420 instrument (Micromeritics Instrument Corp., Norcross, GA, USA). (**a**) Langmuir surface area plot; (**b**) BJH desorption dV/dlog(D) pore volume.

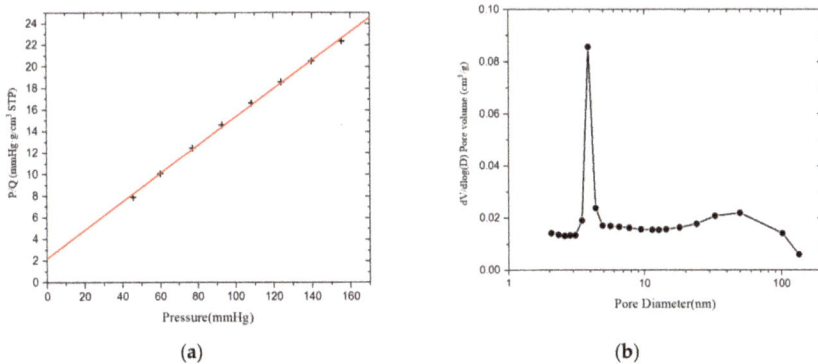

Figure A4. Nitrogen adsorption of sample 2 measured by an ASAP 2420 instrument (Micromeritics). (**a**) Langmuir surface area plot. (**b**) BJH desorption dV/dlog(D) pore volume.

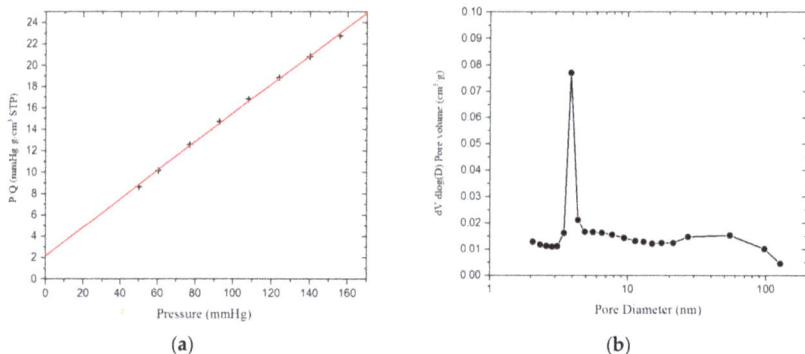

Figure A5. Nitrogen adsorption of sample 3 measured by an ASAP 2420 instrument (Micromeritics). (**a**) Langmuir surface area plot; (**b**) BJH desorption dV/dlog(D) pore volume.

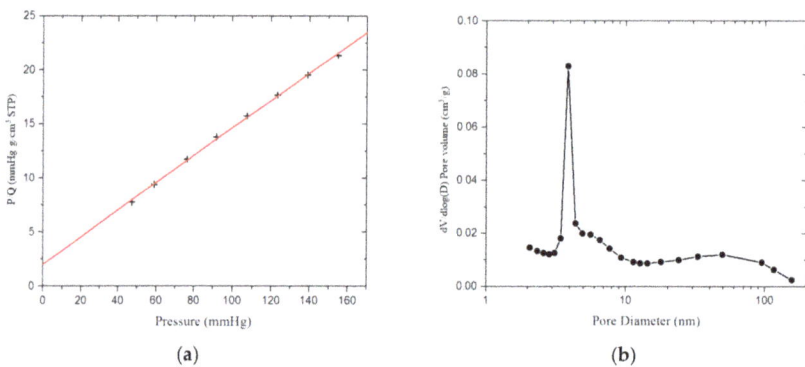

Figure A6. Nitrogen adsorption of sample 4 measured by an ASAP 2420 instrument (Micromeritics). (**a**) Langmuir surface area plot; (**b**) BJH desorption dV/dlog(D) pore volume.

References

1. Yang, Y.; Yang, H.; Tao, L.; Yao, J.; Wang, W.; Zhang, K.; Luquot, L. Microscopic Determination of Remaining Oil Distribution in Sandstones with Different Permeability Scales Using Computed Tomography Scanning. *J. Energy Resour. Technol.* **2019**, *141*, 092903. [CrossRef]
2. Cai, J.; Luo, L.; Ye, R.; Zeng, X.; Hu, X. Recent advances on fractal modeling of permeability for fibrous porous media. *Fractals* **2015**, *23*, 1540006. [CrossRef]
3. Bustin, R.M.; Bustin, A.M.; Cui, A.; Ross, D.; Pathi, V.M. Impact of shale properties on pore structure and storage characteristics. In Proceedings of the SPE Shale Gas Production Conference, Fort Worth, TX, USA, 16–18 November 2008; Society of Petroleum Engineers: Richardson, TX, USA, 2008. [CrossRef]
4. Tang, X.; Jiang, Z.; Li, Z.; Gao, Z.; Bai, Y.; Zhao, S.; Feng, J. The effect of the variation in material composition on the heterogeneous pore structure of high-maturity shale of the Silurian Longmaxi formation in the southeastern Sichuan Basin, China. *J. Nat. Gas Sci. Eng.* **2015**, *23*, 464–473. [CrossRef]
5. Loucks, R.G.; Reed, R.M.; Ruppel, S.C.; Jarvie, D.M. Morphology, genesis, and distribution of nanometer-scale pores in siliceous mudstones of the Mississippian Barnett Shale. *J. Sediment. Res.* **2009**, *79*, 848–861. [CrossRef]
6. Zhou, S.; Yan, G.; Xue, H.; Guo, W.; Li, X. 2D and 3D nanopore characterization of gas shale in Longmaxi formation based on FIB-SEM. *Mar. Pet. Geol.* **2016**, *73*, 174–180. [CrossRef]
7. Loucks, R.G.; Reed, R.M.; Ruppel, S.C.; Hammes, U. Spectrum of pore types and networks in mudrocks and a descriptive classification for matrix-related mudrock pores. *Aapg Bull.* **2012**, *96*, 1071–1098. [CrossRef]

8. Engelder, T.; Cathles, L.M.; Bryndzia, L.T. The fate of residual treatment water in gas shale. *J. Unconv. Oil Gas Resour.* **2014**, *7*, 33–48. [CrossRef]

9. Darabi, H.; Ettehad, A.; Javadpour, F.; Sepehrnoori, K. Gas flow in ultra-tight shale strata. *J. Fluid Mech.* **2012**, *710*, 641–658. [CrossRef]

10. Chalmers, G.R.; Ross, D.J.; Bustin, R.M. Geological controls on matrix permeability of Devonian Gas Shales in the Horn River and Liard basins, northeastern British Columbia, Canada. *Int. J. Coal Geol.* **2012**, *103*, 120–131. [CrossRef]

11. Tinni, A.; Fathi, E.; Agarwal, R.; Sondergeld, C.H.; Akkutlu, I.Y.; Rai, C.S. Shale permeability measurements on plugs and crushed samples. In Proceedings of the SPE Canadian Unconventional Resources Conference, Calgary, AB, Canada, 30 October–1 November 2012; Society of Petroleum Engineers: Richardson, TX, USA, 2012. [CrossRef]

12. Firouzi, M.; Alnoaimi, K.; Kovscek, A.; Wilcox, J. Klinkenberg effect on predicting and measuring helium permeability in gas shales. *Int. J. Coal Geol.* **2014**, *123*, 62–68. [CrossRef]

13. Kim, C.; Jang, H.; Lee, J. Experimental investigation on the characteristics of gas diffusion in shale gas reservoir using porosity and permeability of nanopore scale. *J. Pet. Sci. Eng.* **2015**, *133*, 226–237. [CrossRef]

14. Wang, F.P.; Reed, R.M. Pore networks and fluid flow in gas shales. In Proceedings of the SPE Annual Technical Conference and Exhibition, New Orleans, LA, USA, 4–7 October 2009; Society of Petroleum Engineers: Richardson, TX, USA, 2009. [CrossRef]

15. Keelan, D.; Koepf, E. The role of cores and core analysis in evaluation of formation damage. *J. Pet. Technol.* **1977**, *29*, 482–490. [CrossRef]

16. Shen, Y.; Ge, H.; Meng, M.; Jiang, Z.; Yang, X. Effect of water imbition on shale permeability and its influence on gas production. *Energy Fuels* **2017**, *31*, 4973–4980. [CrossRef]

17. Ghanizadeh, A.; Gasparik, M.; Amann-Hildenbrand, A.; Gensterblum, Y.; Krooss, B. Lithological controls on matrix permeability of organic-rich shales: An experimental study. *Energy Procedia* **2013**, *40*, 127–136. [CrossRef]

18. Hsieh, P.A.; Tracy, J.V.; Neuzil, C.E.; Bredehoeft, J.D.; Silliman, S.E. A transient laboratory method for determining the hydraulic properties of "tight" rocks—I. Theory. *Int. J. Rock Mech. Min. Sci. Geomech. Abstr.* **1981**, *18*, 245–252. [CrossRef]

19. Neuzil, C.E.; Cooley, C.; Silliman, S.E.; Bredehoeft, J.D.; Hsieh, P.A. A transient laboratory method for determining the hydraulic properties of "tight" rocks—II. Application. *Int. J. Rock Mech. Min. Sci. Geomech. Abstr.* **1981**, *18*, 253–258. [CrossRef]

20. Dicker, A.; Smits, R. A practical approach for determining permeability from laboratory pressure-pulse decay measurements. In Proceedings of the International Meeting on Petroleum Engineering, Tianjin, China, 1–4 November 1988; Society of Petroleum Engineers: Richardson, TX, USA, 1988. [CrossRef]

21. Luffel, D.; Hopkins, C.; Schettler, P., Jr. Matrix permeability measurement of gas productive shales. In Proceedings of the SPE Annual Technical Conference and Exhibition, Houston, TX, USA, 3–6 October 1993; Society of Petroleum Engineers: Richardson, TX, USA, 1993. [CrossRef]

22. Jones, S. A technique for faster pulse-decay permeability measurements in tight rocks. *SPE Form. Eval.* **1997**, *12*, 19–26. [CrossRef]

23. Wu, Y.S.; Pruess, K.J. Gas flow in porous media with Klinkenberg effects. *Transp. Porous Media* **1998**, *32*, 117–137. [CrossRef]

24. Cui, X.; Bustin, A.; Bustin, R.M. Measurements of gas permeability and diffusivity of tight reservoir rocks: Different approaches and their applications. *Geofluids* **2009**, *9*, 208–223. [CrossRef]

25. Barral, C.; Oxarango, L.; Pierson, P. Characterizing the gas permeability of natural and synthetic materials. *Transp. Porous Media* **2010**, *81*, 277–293. [CrossRef]

26. Fisher, Q.; Lorinczi, P.; Grattoni, C.; Rybalcenko, K.; Crook, A.J.; Allshorn, S.; Burns, A.D.; Shafagh, I. Laboratory characterization of the porosity and permeability of gas shales using the crushed shale method: Insights from experiments and numerical modelling. *Mar. Pet. Geol.* **2017**, *86*, 95–110. [CrossRef]

27. Profice, S.; Lasseux, D.; Jannot, Y.; Jebara, N.; Hamon, G. Permeability, porosity and klinkenberg coefficient determination on crushed porous media. *Petrophysics* **2012**, *53*, 430–438. Available online: https://www.onepetro.org/journal-paper/SPWLA-2012-v53n6a5 (accessed on 21 March 2019).

28. Klinkenberg, L. The permeability of porous media to liquids and gases. In Proceedings of the Drilling and Production Practice, New York, NY, USA, 1 January 1914; American Petroleum Institute: Washington, DC, USA, 1941. Available online: https://www.onepetro.org/conference-paper/API-41-200 (accessed on 21 March 2019).

29. Florence, F.A.; Rushing, J.; Newsham, K.E.; Blasingame, T.A. Improved permeability prediction relations for low permeability sands. In Proceedings of the Rocky Mountain Oil & Gas Technology Symposium, Denver, CO, USA, 16–18 April 2007; Society of Petroleum Engineers: Richardson, TX, USA, 2007. [CrossRef]

30. Sampath, K.; Keighin, C.W. Factors affecting gas slippage in tight sandstones of cretaceous age in the Uinta basin. *J. Pet. Technol.* **1982**, *34*, 715–720. [CrossRef]

31. Xu, R.N.; Jiang, P.X. Numerical simulation of fluid flow in microporous media. *Int. J. Heat Fluid Flow* **2008**, *29*, 1447–1455. [CrossRef]

32. Sun, Z.; Kang, Y.; Zhang, J. Density functional study of pressure profile for hard-sphere fluids confined in a nano-cavity. *AIP Adv.* **2014**, *4*, 031308. [CrossRef]

33. Lee, J.W.; Nilson, R.H.; Templeton, J.A.; Griffiths, S.K.; Kung, A.; Wong, B.M. Comparison of molecular dynamics with classical density functional and poisson–boltzmann theories of the electric double layer in nanochannels. *J. Chem. Theory Comput.* **2012**, *8*, 2012–2022. [CrossRef]

34. Zeng, K.; Jiang, P.; Lun, Z.; Xu, R. Molecular Simulation of Carbon Dioxide and Methane Adsorption in Shale Organic Nanopores. *Energy Fuels* **2019**, *33*. [CrossRef]

35. Tinni, A.; Odusina, E.; Sulucarnain, I.; Sondergeld, C.; Rai, C. NMR response of brine, oil and methane in organic rich shales. In Proceedings of the SPE Unconventional Resources Conference, The Woodlands, TX, USA, 1–3 April 2014; Society of Petroleum Engineers: Richardson, TX, USA, 2014. [CrossRef]

36. Viswanathan, K.; Kausik, R.; Cao Minh, C.; Zielinski, L.; Vissapragada, B.; Akkurt, R.; Song, Y.-Q.; Liu, C.; Jones, S.; Blair, E. Characterization of gas dynamics in kerogen nanopores by NMR. In Proceedings of the SPE Annual Technical Conference and Exhibition, Denver, CO, USA, 30 October–2 November 2011; Society of Petroleum Engineers: Richardson, TX, USA, 2011. [CrossRef]

37. Odusina, E.; Sigal, R.F. Laboratory NMR Measurements on Methane Saturated Barnett Shale Samples. *Petrophysics* **2011**, *52*, 32–49. Available online: https://www.onepetro.org/journal-paper/SPWLA-2011-v52n1a2 (accessed on 21 March 2019).

38. Achang, M.; Pashin, J.C.; Cui, X. The influence of particle size, microfractures, and pressure decay on measuring the permeability of crushed shale samples. *Int. J. Coal Geol.* **2017**, *183*, 174–187. [CrossRef]

39. Etminan, S.R.; Javadpour, F.; Maini, B.B.; Chen, Z. Measurement of gas storage processes in shale and of the molecular diffusion coefficient in kerogen. *Int. J. Coal Geol.* **2014**, *123*, 10–19. [CrossRef]

40. Gensterblum, Y.; Ghanizadeh, A.; Cuss, R.J.; Amann-Hildenbrand, A.; Krooss, B.M.; Clarkson, C.R.; Harrington, J.F.; Zoback, M.D. Gas transport and storage capacity in shale gas reservoirs—A review. Part A: Transport processes. *J. Unconv. Oil Gas Resour.* **2015**, *12*, 87–122. [CrossRef]

41. Sondergeld, C.H.; Newsham, K.E.; Comisky, J.T.; Rice, M.C.; Rai, C.S. Petrophysical considerations in evaluating and producing shale gas resources. In Proceedings of the SPE Unconventional Gas Conference, Pittsburgh, PA, USA, 23–25 February 2010; Society of Petroleum Engineers: Richardson, TX, USA, 2010. [CrossRef]

42. Heller, R.; Vermylen, J.; Zoback, M. Experimental investigation of matrix permeability of gas shales. *AAPG Bull.* **2014**, *98*, 975–995. [CrossRef]

43. Reis, J.C. Effect of fracture spacing distribution on pressure transient response in naturally fractured reservoirs. *J. Pet. Sci. Eng.* **1998**, *20*, 31–47. [CrossRef]

44. Reeves, S.; Pekot, L. Advanced reservoir modeling in desorption-controlled reservoirs. In Proceedings of the SPE Rocky Mountain Petroleum Technology Conference, Keystone, CO, USA, 21–23 May 2001; Society of Petroleum Engineers: Richardson, TX, USA, 2001. [CrossRef]

45. Zhou, B.; Jiang, P.; Xu, R.; Ouyang, X. General slip regime permeability model for gas flow through porous media. *Phys. Fluids* **2016**, *28*, 072003. [CrossRef]

materials

MDPI

Article

Colorimetric Detection of Mercury Ions in Water with Capped Silver Nanoprisms

Fouzia Tanvir [1], Atif Yaqub [1], Shazia Tanvir [2], Ran An [2] and William A. Anderson [2,*]

[1] Department of Zoology, Government College University, Lahore 54000, Pakistan;
    tanvir.fouzia@gmail.com (F.T.); atifravian@gmail.com (A.Y.)
[2] Department of Chemical Engineering, University of Waterloo, Waterloo, ON N2L 3G1, Canada;
    shazia@genemis.ca (S.T.); r6an@uwaterloo.ca (R.A.)
*   Correspondence: wanderson@uwaterloo.ca

Received: 29 March 2019; Accepted: 7 May 2019; Published: 10 May 2019

Abstract: The emission of mercury (II) from coal combustion and other industrial processes may have impacts on water resources, and the detection with sensitive but rapid testing methods is desirable for environmental screening. Towards this end, silver nanoprisms were chemically synthesized resulting in a blue reagent solution that transitioned towards red and yellow solutions when exposed to Hg^{2+} ions at concentrations from 0.5 to 100 µM. A galvanic reduction of Hg^{2+} onto the surfaces is apparently responsible for a change in nanoprism shape towards spherical nanoparticles, leading to the change in solution color. There were no interferences by other tested mono- and divalent metal cations in solution and pH had minimal influence in the range of 6.5 to 9.8. The silver nanoprism reagent provided a detection limit of approximately 1.5 µM (300 µg/L) for mercury (II), which compared reasonably well with other reported nanoparticle-based techniques. Further optimization may reduce this detection limit, but matrix effects in realistic water samples require further investigation and amelioration.

Keywords: nanoparticles; nanoplates; spectral blue shift; amalgam; water quality

---

1. Introduction

Mercury has been emitted by various industrial processes over the past century, with artisanal gold mining and coal combustion as the two largest sources currently, at 775 and 558 Mg per year respectively [1]. For coal combustion, just under 30% of these emissions are in the divalent form [1], Hg^{2+}, which tends to have a high water solubility and shorter lifetime in the environment resulting in more local deposition [2]. As a monitoring tool, it is desirable to measure mercury content in aqueous samples to identify water sources that may be impacted by mercury emissions, since the US EPA limit for drinking water is 0.01 µM (2 ppb) [3]. The most common laboratory methods, using cold vapor atomic fluorescence spectrometry or atomic absorption spectrometry, are sensitive and accurate but require specialized facilities and equipment [4]. For screening purposes, it is desirable to have simple and rapid measurement techniques that can be used in the field, even if their sensitivity, accuracy and selectivity may not be as good as laboratory instrumentation. Therefore, a variety of other mercury detection methods have been developed over the years, ranging from colorimetric to electrochemical methods [5]. Molecular-based optical methods for heavy metal ion detection [6] have been developed and extended to mercury, including a wide range of chemical-binding and fluorescence detection schemes [7]. Other examples for the detection of aqueous phase mercury (II) include a nanoparticle-functionalized carbon paper electrode [8], a gold nanoparticle-aptamer colorimetric method [9], and a gold nanozyme paper chip [10].

Nanoparticle (NP)-based detection of aqueous phase cations has been targeted for a range of metal analytes, including nickel using glutathione-functionalized gold [11] or silver [12] NPs, lead using glutathione-functionalized gold nanostars [13], and cobalt using glutathione-modified silver NP spherical, plate and rod shapes [14]. The detection mechanism in these reports has been based on the induction of NP aggregation by the cations, resulting in a color or spectral change in the solution. Likewise, DNA-functionalized gold NPs have been shown to detect Hg^{2+} [15] through a similar aggregation mechanism.

Indeed, a variety of gold and silver nanoparticle applications for colorimetric Hg^{2+} detection has been a focus of research in recent years [16]. The aggregation or disaggregation of gold nanoparticles (AuNPs), functionalized or capped with various surface ligands, has been used to generate a spectral shift or color change in the presence of Hg^{2+} [16]. Silver nanoparticles (AgNPs) can likewise undergo a spectral shift due to aggregation, but the reduction of Hg^{2+} by Ag oxidation is also cited as a mechanism for color change [16]. A color change with silver nanoprisms, capped with 1-dodecanethiol, was induced by Hg^{2+} in the presence of added iodide anions and attributed to a morphological transformation in the nanoprisms [17]. Limits of detection for the nanoparticle-based methods vary widely depending on the specific formulation and measurement method (e.g. instrumental versus visual), but they tend to fall within the range of 10 nM to 55 μM mercury (II) ions for AgNPs [16].

A variety of different AgNP synthesis methods have been reported in the context of mercury (II) ion detection. Biological "green" synthesis has been used for AgNPs that shift in colour from yellow to colorless in the presence of mercury ions [18]. Some AgNP detection methods for mercury ion have employed various surface modifications including oligonucleotides [19], glutathione [20], and leaf extracts [21], for example. Chemically-synthesized AgNPs with a cytosine triphosphate cap demonstrated a similar decrease in yellow colour in the presence of mercury [22]. Another chemically-synthesized AgNP with a gelatin functionalization likewise showed a color change from yellow to colorless for Hg^{2+} concentrations as low as 25 nM [23]. A different color change, from orange to yellow, was created using AgNPs aggregated by the combination of mercury and lysine [24].

The majority of reports have been based on spherical AgNPs, and the blue-shift in absorbance generally results in the disappearance of the initial yellow color. A few researchers have used non-spherical nanoparticles such as one using triangular nanoplates assembled into thin films which resulted in a blue color [25]. A nanoprism formulation had a Hg^{2+} detection limit as low as 3.3 nM but required the presence of added iodide [17]. For visual detection, there are some advantages of having a more intense starting color such as the blue color attributed to silver nanoplates and nanoprisms, since the blue-shift caused by mercury may result in a more visible color change from blue to yellow or colorless.

Therefore, in this work, a simple chemical synthesis technique was adopted from earlier work with silver nanoprisms [26] to formulate non-spherical AgNPs with a visually-intense blue starting color. A straightforward capping method using polyvinylpyrrolidone was used to stabilize the AgNPs in solution in a form that may be more commercially suitable for shelf-life (versus DNA and similar functionalizations), and that avoided the need for other reagents such as iodide. The response of these AgNPs was tested to measure the spectral response to Hg^{2+} in water, and the sensitivity and response to other metal cations were determined.

2. Materials and Methods

Analytical grade (or of the highest purity available) chemicals were used. Solutions were prepared with ultra-pure water of typical resistivity 18.2 MΩ·cm. All chemicals were purchased from Sigma Aldrich (Canada) including sodium borohydride ($NaBH_4$, 99.99%), hydrogen peroxide (H_2O_2 30%), silver nitrate ($AgNO_3$, 99.99%), trisodium citrate dihydrate ($C_6H_5O_7Na_3$ $2H_2O$, 99.99%), HNO_3 and polyvinylpyrrolidone (PVP, Mw = 40,000), and the metal salts $BaCl_2$, $CdCl_2$, $Cd(NO_3)_2$, $CoCl_2·6H_2O$, $CuSO_4$, $FeCl_2$, $HgCl_2$, KCl, K_2CrO_4, $MgSO_4$, $MnSO_4$, $NaCl$ and $Pb(NO_3)_2$. The metal ion stock solutions were prepared by dissolving a measured amount of salts in 100 mL deionized water and diluting further as necessary.

Silver nanoprisms were synthesized as follows. Sodium citrate (5.5 mM) was prepared in 100 mL deionized water, followed by adding 340 µL of AgNO$_3$ (30 mM) and 560 µL H$_2$O$_2$ (30%). Then 2.3 mL NaBH$_4$ (100 mM) was added and vigorously stirred. After 2 min, the colorless solution turned yellow and then rapidly darkened until a stable blue color was developed after 5 min. PVP was added at the final concentration of 0.03% as a capping agent to further stabilize the silver nanoprisms in solution. The synthesized AgNPs were stored in the dark and used as a stock solution [27]. The silver nanoprism solution absorbance was set by dilution to approximately 0.6 at 665 nm for use in mercury detection.

The synthesized silver nanoprisms were characterized using UV-visible spectrophotometry (HP 8542 Diode Array, Agilent Technologies, Santa Clara, CA, USA), energy-dispersive spectroscopy (EDX), transmission electron microscopy (TEM, Philips CM10, Amsterdam, The Netherlands), dynamic light scattering (DLS) and zeta potential. The colloidal solutions were centrifuged at 10,000 g and washed three times with deionized water. The washed samples were prepared by drying nanoparticles on a carbon tape. The elemental analysis of the nanoparticles was performed by energy dispersive X-ray (EDX) attached to the scanning electron microscope (model FEI/Philips XL30 FEG ESEM, Amsterdam, The Netherlands). The TEM samples were prepared by drop-coating the aqueous solution of nanoprisms onto a carbon coated copper grid (200 mesh), followed by air-drying for 2 h. TEM characterization was performed using a Philips CM10.

A Zetasizer Nano ZS90 (Malvern Instruments Ltd, Malvern, UK) was used to measure particle size distribution by DLS, polydispersity index, and zeta potential, using triplicate runs with 10 measurements in each, and the Dispersion Technology software 5.1. For this analysis, a refractive index of 1.5 was used and the viscosity was assumed to be equal to that of the dispersant liquid.

The PVP capped nanoprisms were tested with various metal ions in aqueous solution with pH ranging from 5.8 to 9.7. The pH-dependent response was tested by dissolving the previously mentioned heavy metal salts in buffer (5 mM phosphate buffered saline). For the detection of Hg^{2+}, solutions of different concentrations were mixed with the AgNP reagent in a 1:1 ratio in a buffer solution and left at room temperature for 30 min, after which the absorption spectra were measured. The comparison of the nanoparticle response to Hg^{2+} versus other metal ions (Ba^{2+}, Cd^{2+}, Co^{2+}, Cu^{2+}, Fe^{2+}, K$^+$, Mg^{2+}, Mn^{2+}, Na$^+$, Pb^{2+}, Zn^{2+}) was investigated under the same conditions. EDTA (ethylenediaminetetraacetic acid) chelation was used as a negative control to confirm the reaction with mercury (II) by binding the metal ions in solution. The mercury ions were mixed with 10 mM of EDTA solution before the addition of the AgNPs.

3. Results and Discussion

3.1. Characterization of the Nanoparticles

Silver nanoprisms were prepared as described using the chemical reduction of silver nitrate to form nanospheres and a yellowish color in solution. This is followed by the transition of the solution to a blue color as the hydrogen peroxide induces anisotropic oxidation of portions of the nanoparticles into Ag$^+$ ions, with a concurrent transformation of the spheres into nanoprisms [28].

To verify the nature of the prepared colloidal suspension, the surface plasmon resonance (SPR) peak is seen in the absorbance spectrum (Figure 1). The spectral measurements of the colloidal solution clearly reflect a mixture of shapes since it features a small shoulder at around 450–475 nm (typical for nanospheres, with out-of-plane dipole resonance), and a more intense band at around 665 nm which has been reported previously for nanoprisms [26] with the in-plane dipole resonance of flatter shapes [29].

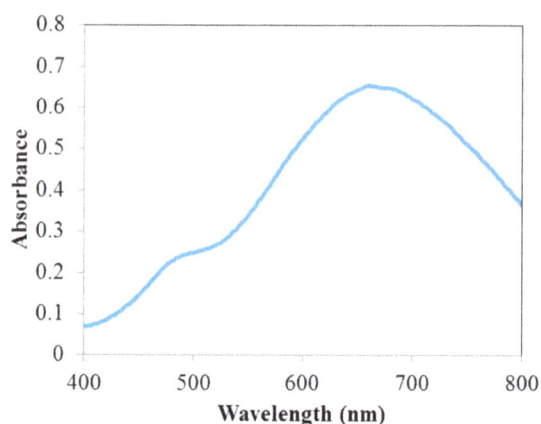

Figure 1. Absorbance spectrum of the as-prepared silver nanoparticle (AgNP) reagent solution before exposure to mercury (II) ion.

The morphology suggested by the spectral measurement is verified by TEM imaging, where the AgNPs (Figure 2A) appear as a mixture of larger nanoprisms with a size of approximately 20 to 50 nm, and some smaller nanospheres, as was also shown in prior work using a similar synthesis method [26]. The particle size distribution profile was determined by DLS (Figure 2B), with an average size of 34.5 ± 6 nm, which was in reasonable agreement with the TEM results shown here and in other work. The polydispersity index reported by the instrument was lower than 0.3 in all cases, which is indicative of a relatively monodisperse system. The particles were determined to be negatively charged with a zeta potential of −27.6 ±2 mV (Figure 3).

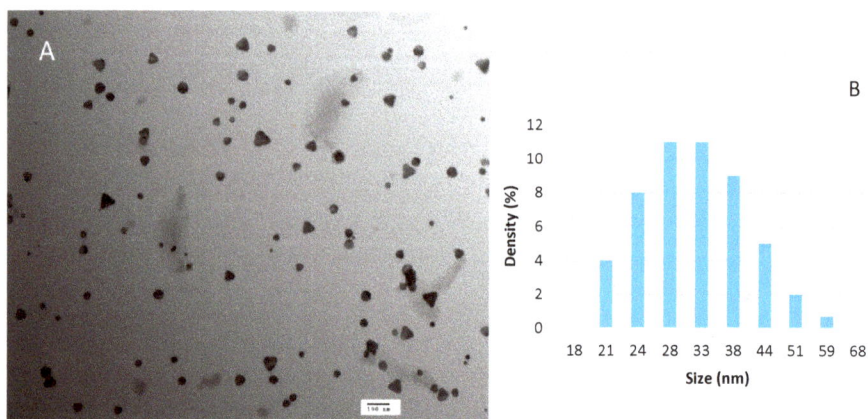

Figure 2. (**A**) TEM micrograph with 100 nm scale bar, and (**B**) particle size distribution profile of silver nanoparticles measured by dynamic light scattering (DLS).

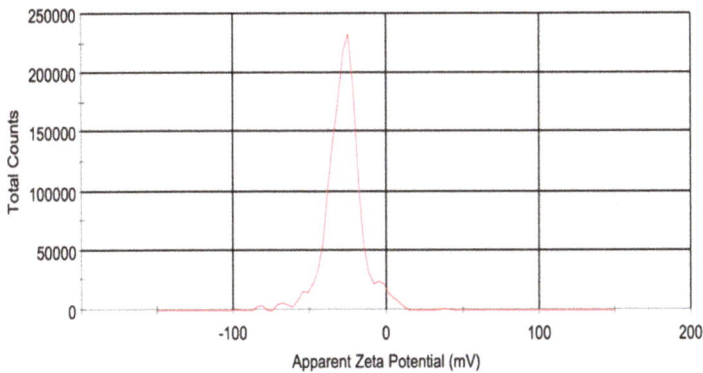

Figure 3. Surface charge distribution of Ag nanoprisms as measured by zeta potential.

Elemental analysis of the AgNPs was performed using EDX (energy-dispersive X-ray spectroscope). The EDX spectrum (not shown) clearly indicated that the prepared samples were pure silver with no contaminating substances other than a peak corresponding to carbon, which can be attributed to the carbon-coated tape sample substrate, and possibly the PVP capping agent.

3.2. Spectral Shifts in the Presence of Hg+

The AgNPs were used to detect Hg^{2+} using the shift of the maximum absorption wavelength after a 30 min incubation time. As shown in Figure 4, the AgNPs in the absence of mercury were characterized by a blue color and peak at 665 nm (curve 1). When Hg^{2+} was added (curves 2 and 3) the peak absorbance wavelength blue-shifted to shorter wavelengths, resulting in a more reddish visual color (see Figure 9 for color photograph).

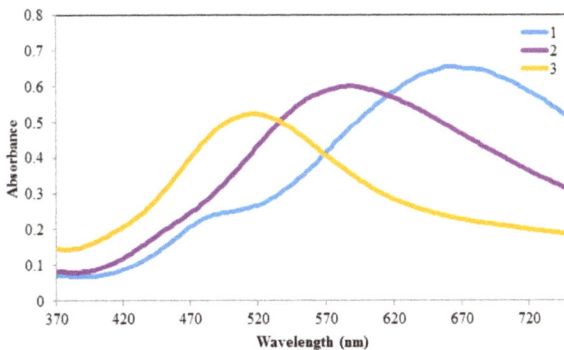

Figure 4. Absorbance spectra of the AgNPs in the absence (1, right) and presence of 2.5 μM (2, center) and 5 μM (3, left) $HgCl_2$, after exposure for 30 min.

Thus, a blue-shift of the absorbance peak was observed in these experiments with a color change from blue (peak around 650 nm) to purple (peak around 530 nm) and eventually to yellow (peak around 450 nm) at high mercury concentrations. Some previous work has attributed the color shift to conformational changes in surface ligands, and/or aggregation of nanoparticles [15,19]. Other work has noted the reduction of Hg^{2+} onto the silver in the presence of H_2O_2 [20], and morphological changes to nanoprisms in the presence of iodide [17].

In this work, there was some evidence for aggregation, as indicated in the TEM images in Figure 5. However, the induction of aggregation by the divalent mercury cation is difficult to reconcile with the lack of response to other divalent cations, as discussed below. However, as can also be seen in Figure 5, the corners of some of the prismatic AgNPs were rounded or possibly removed upon interaction with Hg^{2+}, similar to morphological changes noted by others [17]. This resulted in a mixture of shapes, as illustrated in Figure 5, with relatively fewer and smaller prismatic shapes as compared to the image in Figure 2. These morphological changes and the lack of response to other cations (discussed below) suggest that aggregation did not play the only role in the spectral response and color change.

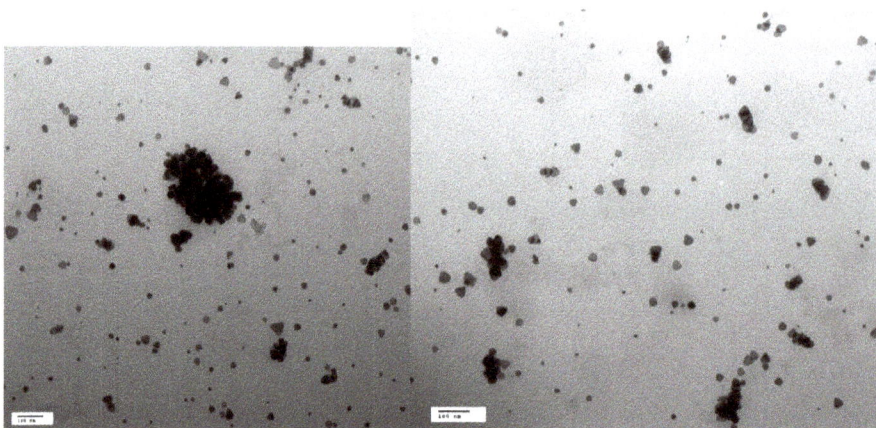

Figure 5. TEM images, with 100 nm scale bar, of AgNPs after reaction with 10 μM Hg^{2+} for 30 min.

The sharp edges of the nanoprisms are more likely prone to attack and in the presence of Hg^{2+} electrons could be initially extracted from the "corner" areas of the Ag nanoprisms resulting in the shape transformation. Unlike other reports of nanoprism morphology changes induced by Hg^{2+} [17], the response measured here did not require the addition of iodide and a thiol. The combination of silver or gold with mercury can lead to bimetallic colloids or amalgam formation, and this also leads to a blue shift of the absorption peak [30,31], although one report shows a slight red-shift in the presence of hydrogen peroxide [32]. An analysis of the AgNPs after exposure to mercury ion performed by EDX confirmed the presence of Hg in the colloids (Figure 6). Here, the Hg^{2+} detection appears to also be achieved by forming an Ag/Hg amalgam upon reduction of the mercury ion to elemental mercury by silver. The PVP AgNP capping agent may act as an electron donor, and the mercury reduction might also be supported by the PVP reduction abilities [33]. However, in the presence of the strong oxidizing agent (H_2O_2) used in the preparation of the nanoprisms, the continued presence of active reducing agents seems uncertain. Therefore, a galvanic replacement reaction, whereby a more reactive metal is dissolved and replaced by a less reactive one, seems more likely and this can occur rapidly at the nanoscale [34].

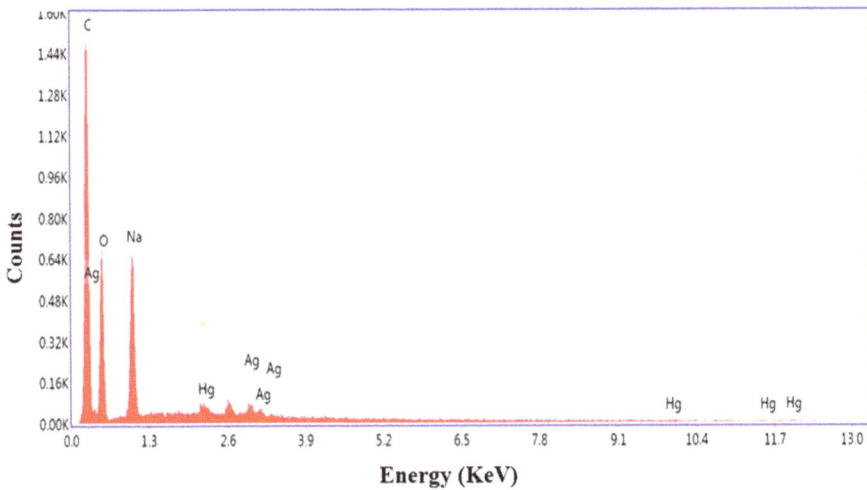

Figure 6. Elemental analysis, performed by energy dispersive X-ray spectroscopy (EDX), of Ag nanoparticles after exposure to mercury, indicating the formation of Ag-Hg amalgams.

3.3. Sensitivity and Selectivity

The effect of pH on mercury detection was tested with 5 μM Hg^{2+} and the extent of the absorbance wavelength peak shift was quantified. As shown in Figure 7, the response of the AgNPs was relatively insensitive to pH ranging from 6.5 to 9.79, confirming that the PVP capped AgNPs exhibited excellent stability towards changes in pH in the neutral range [35]. A pH of 7.2 was selected to subsequently determine the sensitivity and selectivity since a commercialized test kit could readily incorporate a pH adjustment to neutral pH if necessary.

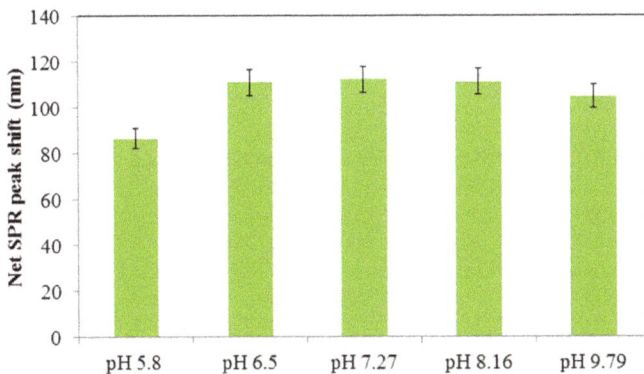

Figure 7. Influence of pH on the extent of the SPR peak shift at 5 μM of Hg^{2+} in 5 mM phosphate buffered saline.

To evaluate the sensitivity of the nanoprisms to Hg^{2+} concentration, the absorption spectra of the AgNP solutions were measured under the optimized conditions, using different Hg^{2+} concentrations added to the solution, with incubation for 30 min. Figure 8A shows that increasing concentrations of Hg^{2+} resulted in a larger blue shift of the absorbance peak, as attributed to the morphology changes illustrated earlier. Therefore, the extent of the peak shift could potentially be employed as a quantitative analysis for Hg^{2+}. This change in peak wavelength versus Hg^{2+} concentration (0.5–100 µM) relationship is shown in Figure 8B. A linear relationship was obtained over the range of 0 to 5 µM with a correlation coefficient of 0.993 (Figure 8B inset). The results indicate that this AgNP material could be used to detect Hg^{2+} at a detection limit of approximately 1.5 µM (approximately 300 µg/L), based on three standard deviations. Although not the most sensitive detection limit ever reported, it compares favorably with the range of 10 nM to 55 µM for other AgNPs reported in one review [16]. The standard deviations shown in Figure 8B suggest that the AgNPs may not be suitable for accurate quantification of Hg^{2+} at lower concentrations, but they may serve the purpose of rapid screening and "yes/no" detection of Hg^{2+} above some minimum threshold. Further improvements in sensitivity may be possible with optimization of the concentrations and ratios of AgNPs to sample solutions.

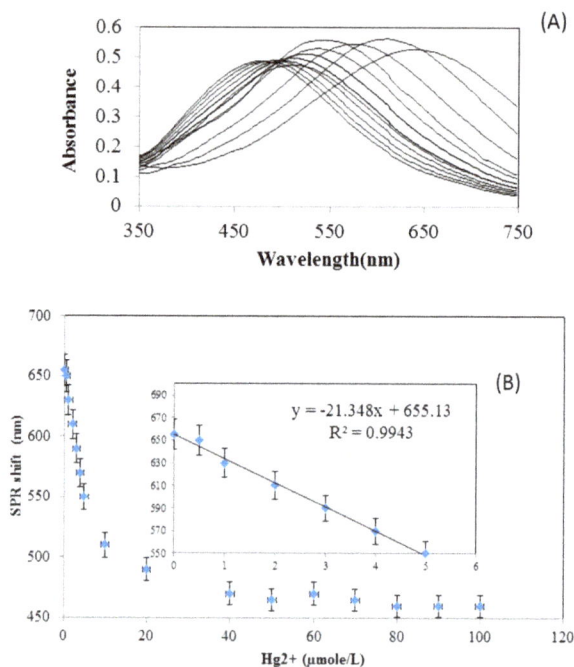

Figure 8. (**A**) UV–visible absorption spectra of the AgNPs after the addition of Hg^{2+} at various concentrations, with the peak shifting from the right (for 0 µM) to left as Hg^{2+} concentration increases up to 100 µM. (**B**) Peak wavelength of the AgNPs at varying concentrations of Hg^{2+} (0–100 µM), where the inset graph shows a linear regression for the range of 0 to 5 µM. The error bars show standard deviations (three measurements).

If a galvanic replacement reaction between Hg^{2+} and Ag^0 is responsible for the detection mechanism and not just aggregation, as discussed earlier, then the AgNPs should not be reactive towards other cations in solution. To determine and verify the response of the nanoprisms towards Hg^{2+}, the effects of Ba^{2+}, Cd^{2+}, Co^{2+}, Cu^{2+}, Fe^{2+}, K^+, Mg^{2+}, Mn^{2+}, Na^+, Pb^{2+}, and Zn^{2+} were determined individually under the selected test conditions. The results shown in Figure 9 indicated that Hg^{2+} alone caused a significant effect on the silver nanoparticles, as determined by spectroscopy (some examples of which color are shown in the Figure 9 inset). A 100-fold higher concentration of the other metal ions also had no significant effect on the SPR peak shift (not shown). These results indicate that the AgNPs exhibited a very high reactivity toward Hg^{2+} ions only, and provide further evidence that the interaction between the AgNPs and Hg^{2+} is primarily driven by a galvanic reaction. The standard reduction potentials for Hg^{2+} and Ag^+ are 0.85 and 0.80 V, respectively, while those of the other species tested here are all <0.80 V [36]. Therefore, of all these metals only Hg^{2+} can potentially be reduced by silver, lending support to the hypothesis that a galvanic reaction is primarily responsible for the nanoprism morphology changes and SPR blue-shift, rather than aggregation effects alone. There are very few, if any, metals of environmental significance with a reduction potential >0.80 V, suggesting that the silver nanoprisms will always be selective towards mercury, in the absence of any other interfering effects that remain to be discovered. Mixtures of the other metal cations and Hg^{2+} were not tested but should be in future work to ensure that surface complexation by other metals will not inhibit or affect the AgNP interactions. Likewise, since these experiments were carried out with $HgCl_2$ only, the effects of other counter-ions such as nitrate would be of interest to explore.

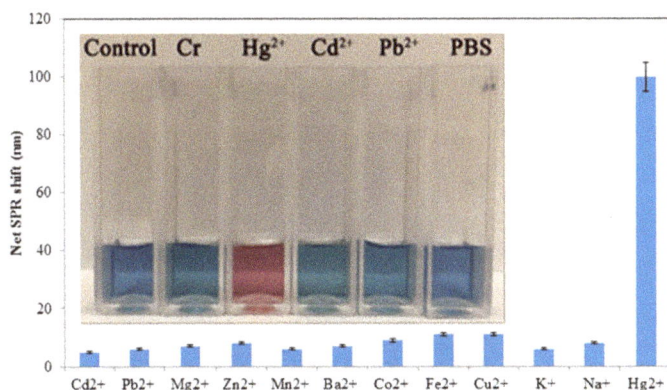

Figure 9. Effect of various metal ions (5 μM) dissolved in 5 mM phosphate buffered saline pH 7.2 on the net SPR shift, showing high reactivity with only Hg^{2+}. Inset: photograph of AgNP solutions, showing color change with Hg^{2+} and lack of significant response to phosphate-buffered saline (PBS), deionized water (Control), and the other ions Cd^{2+}, Pb^{2+} and CrO_4^{2-} (labelled "Cr") at 10 μM.

To further explore the nature of the nanoprism/metal interactions, negative control experiments were performed using the metal ion chelator EDTA. The various metal ions and Hg (II) were mixed with 10 mM EDTA and then allowed to interact with the prismatic silver reagent for 30 min. No color change was noted under these conditions, confirming that the previously measured effects are due to free Hg^{2+} and that chelated ions may not be detectable via this method.

Furthermore, some tests were completed in 0.2 μm filtered municipal tap water to assess the potential interference from other typical water species, with the results shown in Figure 10. Unlike the linear response in pure water and buffer solutions, AgNP responses in the tap water matrix showed a complex non-linear response to increasing Hg^{2+} concentrations. Further work remains to be pursued to determine the reasons behind this difference, but the presence of the oxidant hypochlorite (approximately 0.5 mg/L), alkalinity, or organic carbon such as humic substances may play a role. Matrix effects continue to be a challenge for many nanoparticle-based detection methods [37].

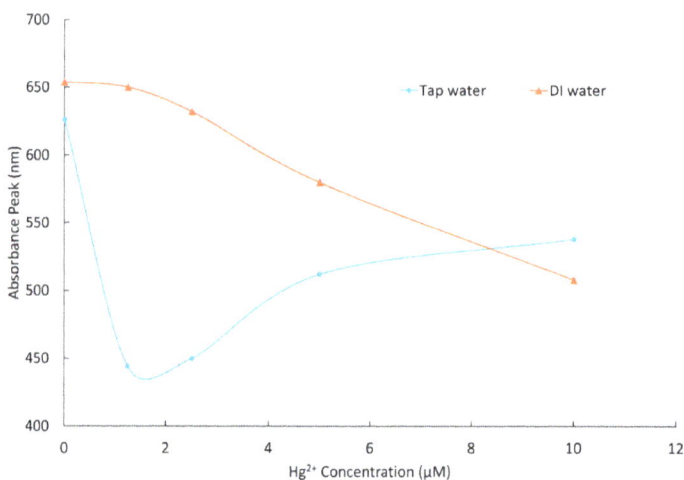

Figure 10. Silver nanoprism solution absorbance peak as a function of Hg^{2+} concentration in deionized (DI) water (upper curve) and municipal tap water (lower curve), showing non-linear blue-shift in the tap water matrix.

4. Conclusions

The technique explored in this study, using PVP-capped prismatic silver nanoparticles, provides a rapid, reasonably sensitive and very specific detection method for aqueous Hg^{2+} samples and may be potentially suitable for remote field and environmental analysis where more advanced instrumentation is not readily available. A galvanic reduction of Hg onto the silver leads to a loss of prismatic size and shape and a blue-shift in the absorbance spectrum, as suggested by TEM, the EDX spectrum, and the lack of sensitivity to other metal ions. The minimum detection limit for this reagent was found to be approximately 1.5 µM under controlled conditions, which may be suitable for environmental screening purposes but not for drinking water testing which requires a 0.01 µM or lower detection limit. The response of the AgNPs to Hg^{2+} alone was confirmed through experiments with a variety of other common metals ions in solution. Interactions with other cations and anions may alter the response to Hg^{2+}, as shown in testing with a municipal water sample and this requires further study to identify the interference mechanisms. Using these prismatic silver nanoparticles may offer a useful approach for the detection of Hg^{2+} in aqueous environmental samples, but additional optimization work is required to lower the detection limit further and to ascertain how to minimize water matrix effects.

Author Contributions: Conceptualization, F.T. and S.T.; methodology, F.T., S.T., and R.A.; investigation, F.T. and R.A.; resources, W.A.A.; writing—original draft preparation, F.T. and R.A.; writing—review and editing, S.T., A.Y. and W.A.A.; supervision, A.Y. and W.A.A.; project administration, W.A.A. and S.T.; funding acquisition, W.A.A.

Funding: This research was funded in part by the Ontario Centers of Excellence Voucher for Innovation and Productivity I program. F. Tanvir was supported by an International Research Support Initiative Program grant offered by the Higher Education Commission (HEC) Islamabad, Pakistan.

Conflicts of Interest: The authors declare no conflict of interest. The funders had no role in the design of the study; in the collection, analyses, or interpretation of data; in the writing of the manuscript, or in the decision to publish the results.

References

1. Streets, D.G.; Horowitz, H.M.; Lu, Z.; Levin, L.; Thackray, C.P.; Sunderland, E.M. Global and regional trends in mercury emissions and concentrations, 2010–2015. *Atmos. Environ.* **2019**, *201*, 417–427. [CrossRef]

2. Zhang, L.; Wang, S.; Wu, Q.; Wang, F.; Lin, C.J.; Zhang, L.; Hui, M.; Yang, M.; Su, H.; Hao, J. Mercury transformation and speciation in flue gases from anthropogenic emission sources: A critical review. *Atmos. Chem. Phys.* **2016**, *16*, 2417–2433. [CrossRef]

3. US EPA. Available online: https://safewater.zendesk.com/hc/en-us/articles/212076077-4-What-are-EPA-s-drinking-water-regulations-for-mercury- (accessed on 18 April 2019).

4. Leopold, K.; Foulkes, M.; Worsfold, P. Methods for the determination and speciation of mercury in natural waters—A review. *Anal. Chim. Acta* **2010**, *663*, 127–138. [CrossRef]

5. Martín-Yerga, D.; González-García, M.B.; Costa-García, A. Electrochemical determination of mercury: A review. *Talanta* **2013**, *116*, 1091–1104. [CrossRef] [PubMed]

6. Uglov, A.N.; Bessmertnykh-Lemeune, A.; Guilard, R.; Averin, A.D.; Beletskaya, I.P. Optical methods for the detection of heavy metal ions. *Russ. Chem. Rev.* **2014**, *83*, 196. [CrossRef]

7. Chen, G.; Guo, Z.; Zeng, G.; Tang, L. Fluorescent and colorimetric sensors for environmental mercury detection. *Analyst* **2015**, *140*, 5400–5443. [CrossRef]

8. Bui, M.P.N.; Brockgreitens, J.; Ahmed, S.; Abbas, A. Dual detection of nitrate and mercury in water using disposable electrochemical sensors. *Biosens. Bioelectron.* **2016**, *85*, 280–286. [CrossRef] [PubMed]

9. Wang, Y.; Yang, F.; Yang, X. Colorimetric biosensing of mercury (II) ion using unmodified gold nanoparticle probes and thrombin-binding aptamer. *Biosens. Bioelectron.* **2010**, *25*, 1994–1998. [CrossRef]

10. Han, K.N.; Choi, J.S.; Kwon, J. Gold nanozyme-based paper chip for colorimetric detection of mercury ions. *Sci. Rep.* **2017**, *7*, 2806. [CrossRef] [PubMed]

11. Fu, R.; Li, J.; Yang, W. Aggregation of glutathione-functionalized Au nanoparticles induced by Ni 2+ ions. *J. Nanopart. Res.* **2012**, *14*, 929. [CrossRef]

12. Li, H.; Cui, Z.; Han, C. Glutathione-stabilized silver nanoparticles as colorimetric sensor for Ni2+ ion. *Sens. Actuators B Chem.* **2009**, *143*, 87–92. [CrossRef]

13. D'Agostino, A.; Taglietti, A.; Bassi, B.; Donà, A.; Pallavicini, P. A naked eye aggregation assay for Pb 2+ detection based on glutathione-coated gold nanostars. *J. Nanopart. Res.* **2014**, *16*, 2683.

14. Sung, H.K.; Oh, S.Y.; Park, C.; Kim, Y. Colorimetric detection of Co2+ ion using silver nanoparticles with spherical, plate, and rod shapes. *Langmuir* **2013**, *29*, 8978–8982. [CrossRef]

15. Lee, J.S.; Han, M.S.; Mirkin, C.A. Colorimetric detection of mercuric ion (Hg2+) in aqueous media using DNA-functionalized gold nanoparticles. *Angew. Chem. Int. Ed.* **2007**, *46*, 4093–4096. [CrossRef] [PubMed]

16. Zarlaida, F.; Adlim, M. Gold and silver nanoparticles and indicator dyes as active agents in colorimetric spot and strip tests for mercury (II) ions: A review. *Microchim. Acta* **2017**, *184*, 45–58. [CrossRef]

17. Chen, L.; Fu, X.; Lu, W.; Chen, L. Highly sensitive and selective colorimetric sensing of Hg2+ based on the morphology transition of silver nanoprisms. *ACS Appl. Mater. Interfaces* **2012**, *5*, 284–290. [CrossRef]

18. Farhadi, K.; Forough, M.; Molaei, R.; Hajizadeh, S.; Rafipour, A. Highly selective Hg2+ colorimetric sensor using green synthesized and unmodified silver nanoparticles. *Sens. Actuators B Chem.* **2012**, *161*, 880–885. [CrossRef]

19. Wang, Y.; Yang, F.; Yang, X. Colorimetric detection of mercury (II) ion using unmodified silver nanoparticles and mercury-specific oligonucleotides. *ACS Appl. Mater. Interfaces* **2010**, *2*, 339–342. [CrossRef] [PubMed]

20. Alam, A.; Ravindran, A.; Chandran, P.; Khan, S.S. Highly selective colorimetric detection and estimation of Hg2+ at nano-molar concentration by silver nanoparticles in the presence of glutathione. *Spectrochim. Acta A* **2015**, *137*, 503–508. [CrossRef] [PubMed]

21. Manivel, P.; Ilanchelian, M. Selective and Sensitive Colorimetric Detection of Hg 2+ at Wide pH Range Using Green Synthesized Silver Nanoparticles as Probe. *J. Clust. Sci.* **2017**, *28*, 1145–1162. [CrossRef]

22. Zhan, L.; Yang, T.; Zhen, S.J.; Huang, C.Z. Cytosine triphosphate-capped silver nanoparticles as a platform for visual and colorimetric determination of mercury (II) and chromium (III). *Microchim. Acta* **2017**, *184*, 3171–3178. [CrossRef]

23. Jeevika, A.; Shankaran, D.R. Functionalized silver nanoparticles probe for visual colorimetric sensing of mercury. *Mater. Res. Bull.* **2016**, *83*, 48–55. [CrossRef]

24. Li, L.; Gui, L.; Li, W. A colorimetric silver nanoparticle-based assay for Hg (II) using lysine as a particle-linking reagent. *Microchim. Acta* **2015**, *182*, 1977–1981. [CrossRef]

25. Detsri, E. Novel colorimetric sensor for mercury (II) based on layer-by-layer assembly of unmodified silver triangular nanoplates. *Chin. Chem. Lett.* **2016**, *27*, 1635–1640. [CrossRef]

26. Tanvir, F.; Yaqub, A.; Tanvir, S.; Anderson, W. Poly-l-arginine coated silver nanoprisms and their anti-bacterial properties. *Nanomaterials* **2017**, *7*, 296. [CrossRef]
27. Panzarasa, G. Just What Is It That Makes Silver Nanoprisms so Different, so Appealing? *J. Chem. Educ.* **2015**, *92*, 1918–1923. [CrossRef]
28. Torres, V.; Popa, M.; Crespo, D.; Moreno, J.M.C. Silver nanoprism coatings on optical glass substrates. *Microelectron. Eng.* **2007**, *84*, 1665–1668. [CrossRef]
29. D'Agostino, A.; Taglietti, A.; Desando, R.; Bini, M.; Patrini, M.; Dacarro, G.; Cucca, L.; Pallavicini, P.; Grisoli, P. Bulk surfaces coated with triangular silver nanoplates: Antibacterial action based on silver release and photo-thermal effect. *Nanomaterials* **2017**, *7*, 7. [CrossRef]
30. Rex, M.; Hernandez, F.E.; Campiglia, A.D. Pushing the limits of mercury sensors with gold nanorods. *Anal. Chem.* **2006**, *78*, 445–451. [CrossRef]
31. Kamali, K.Z.; Pandikumar, A.; Jayabal, S.; Ramaraj, R.; Lim, H.N.; Ong, B.H.; Bien, C.S.D.; Kee, Y.Y.; Huang, N.M. Amalgamation based optical and colorimetric sensing of mercury (II) ions with silver@graphene oxide nanocomposite materials. *Microchim. Acta* **2016**, *183*, 369–377. [CrossRef]
32. Wang, G.L.; Zhu, X.Y.; Jiao, H.J.; Dong, Y.M.; Li, Z.J. Ultrasensitive and dual functional colorimetric sensors for mercury (II) ions and hydrogen peroxide based on catalytic reduction property of silver nanoparticles. *Biosens. Bioelectron.* **2012**, *31*, 337–342. [CrossRef]
33. Wu, C.; Mosher, B.P.; Lyons, K.; Zeng, T. Reducing ability and mechanism for polyvinylpyrrolidone (PVP) in silver nanoparticles synthesis. *J. Nanosci. Nanotechnol.* **2010**, *10*, 2342–2347. [CrossRef]
34. Wang, Y.; Wen, G.; Ye, L.; Liang, A.; Jiang, Z. Label-free SERS study of galvanic replacement reaction on silver nanorod surface and its application to detect trace mercury ion. *Sci. Rep.* **2016**, *6*, 19650. [CrossRef]
35. Badawy, A.M.E.; Luxton, T.P.; Silva, R.G.; Scheckel, K.G.; Suidan, M.T.; Tolaymat, T.M. Impact of environmental conditions (pH, ionic strength, and electrolyte type) on the surface charge and aggregation of silver nanoparticles suspensions. *Environ. Sci. Technol.* **2010**, *44*, 1260–1266. [CrossRef]
36. Haynes, W.M. *The CRC Handbook of Chemistry and Physics*, 97th ed.; CRC Press/Taylor & Francis: Boca Raton, FL, USA, 2017; p. 192+.
37. Metreveli, G.; Philippe, A.; Schaumann, G.E. Disaggregation of silver nanoparticle homoaggregates in a river water matrix. *Sci. Total Environ.* **2015**, *535*, 35–44. [CrossRef]

![materials logo] *materials*

MDPI

Article

Study on Adsorption Mechanism and Failure Characteristics of CO_2 Adsorption by Potassium-Based Adsorbents with Different Supports

Bao-guo Fan, Li Jia, Yan-lin Wang, Rui Zhao, Xue-song Mei, Yan-yan Liu and Yan Jin *

College of Electrical and Power Engineering, Taiyuan University of Technology, Taiyuan 030024, China; fanbaoguo@tsinghua.org.cn (B.-g.F.); 18734869558@163.com (L.J.); 13453122526@163.com (Y.-l.W.); 19834430509@163.com (R.Z.); explorer2018@163.com (X.-s.M.); 18334703500@163.com (Y.-y.L.)
* Correspondence: jinyan@tyut.edu.cn; Tel.: +86-139-3463-0502

Received: 31 October 2018; Accepted: 27 November 2018; Published: 30 November 2018

Abstract: In order to obtain the adsorption mechanism and failure characteristics of CO_2 adsorption by potassium-based adsorbents with different supports, five types of supports (circulating fluidized bed boiler fly ash, pulverized coal boiler fly ash, activated carbon, molecular sieve, and alumina) and three kinds of adsorbents under the modified conditions of K_2CO_3 theoretical loading (10%, 30%, and 50%) were studied. The effect of the reaction temperature (50 °C, 60 °C, 70 °C, 80 °C, and 90 °C) and CO_2 concentration (5%, 7.5%, 10%, 12.5%, and 15%) on the adsorption of CO_2 by the adsorbent after loading and the effect of flue gas composition on the failure characteristics of adsorbents were obtained. At the same time, the microscopic characteristics of the adsorbents before and after loading and the reaction were studied by using a specific surface area and porosity analyzer as well as a scanning electron microscope and X-ray diffractometer. Combining its reaction and adsorption kinetics process, the mechanism of influence was explored. The results show that the optimal theoretical loading of the five adsorbents is 30% and the reaction temperature of 70 °C and the concentration of 12.5% CO_2 are the best reaction conditions. The actual loading and CO_2 adsorption performance of the K_2CO_3/AC adsorbent are the best while the K_2CO_3/Al_2O_3 adsorbent is the worst. During the carbonation reaction of the adsorbent, the cumulative pore volume plays a more important role in the adsorption process than the specific surface area. As the reaction temperature increases, the internal diffusion resistance increases remarkably. K_2CO_3/AC has the lowest activation energy and the carbonation reaction is the easiest to carry out. SO_2 and HCl react with K_2CO_3 to produce new substances, which leads to the gradual failure of the adsorbents and K_2CO_3/AC has the best cycle failure performance.

Keywords: potassium-based adsorbent; load modification; CO_2 adsorption; failure; kinetics; microscopic characteristics

---

1. Introduction

Global climate change caused by greenhouse gas emissions is a hot issue in our modern society, which is related to the development and survival of the whole mankind. China's carbon emissions account for 29% of the global total and rank first in the world [1]. The current CO_2 emission reduction technologies are mainly divided into four ways: pre-combustion decarbonization [2,3], chemical chain circulation [4,5], pure oxygen combustion [6,7], and post-combustion capture [8–11]. Among them, CO_2 capture and storage (CCS) technology has been widely used [12–15], but its cost is high [16,17]. However, alkali metal carbonates such as Na_2CO_3 and K_2CO_3 have become the promising CO_2 adsorbents due to their low cost, low secondary pollution, and high cycle efficiency [18,19].

It has been found that potassium-based adsorbents can remove CO_2 at a low temperature (60–80 °C) under the conditions in which water vapor is adsorbed [20]. However, its adsorption efficiency is low. Some researchers have modified the potassium-based adsorbent by using activated carbon (AC), MgO, TiO_2 [21], and a 5A molecular sieves [22] and studied the carbonation reaction by thermogravimetric analyzer. They found that the CO_2 adsorption rate of K_2CO_3/AC adsorbent increased by 73% as the AC loading increased from 9% to 33% under the conditions of 60 °C and 10% CO_2 [23]. When the Al_2O_3 loading increased from 12.8% to 36.8%, the CO_2 adsorption rate of the K_2CO_3/Al_2O_3 adsorbent increased by 62% [24]. However, the above studies only use the thermogravimetric analyzer to study the carbonation reaction process by the weight loss of the adsorbent. The mass of the adsorbent involved in the reaction is small and the range of loading is narrow.

Since the potassium-based adsorbent can be regenerated in the temperature range of 120 to 200 °C, it provides the possibility of an absorption-regeneration cycle of the adsorbent, which achieves the efficient removal of CO_2. At the same time, the material composition and microstructure of fly ash in coal-fired power plants are very similar to those of activated carbon. Hence, the use of fly ash as a support can make up for the disadvantages of low utilization efficiency of fly ash alone and also modify the potassium-based adsorbent to improve the removal efficiency of CO_2. However, due to the large difference between the combustion mode of pulverized coal (PC) and circulating fluidized bed boiler (CFB) and the fuel used, the chemical composition and microstructure of the fly ash produced are quite different.

In addition, the reaction and adsorption kinetics studies have become an important method for predicting the adsorption rate-determining step and analyzing the adsorption mechanism. It is widely used in the adsorption of heavy metals from the liquid phase and the adsorption of SO_2 and NO on the surface of solid adsorbents. At present, there are very limited studies on the adsorption kinetics, thermodynamics, and adsorption equilibrium of CO_2 on the surface of adsorbents.

In summary, the adsorption of CO_2 by potassium-based adsorbents is related to its characteristics and the focus of the above studies is scattered. Although there have been studies on the modification of potassium-based adsorbents by different supports, the related effects vary greatly depending on the type of support. Among them, the studies on the modification using fly ash as a support have rarely been reported. The study of adsorption kinetics of CO_2 by adsorbents is also relatively small and the relevant mechanisms are not fully explained. On the basis of different supports' effects on the CO_2 adsorption characteristics of potassium-based adsorbents combined with the microscopic properties of the adsorbent, the mechanism of carbonation is studied using reaction and adsorption kinetics. Moreover, the effect of flue gas composition on the failure characteristics of the adsorbent is studied, which will provide a theoretical basis for future CO_2 removal methods.

2. Research Object and Method

2.1. Preparation and Characterization of Samples

K_2CO_3 (Bodi chemical industry, Tianjin, China) was chosen as the active component of the adsorbent sample in this study and circulating fluidized bed fly ash (CFA) (Ping Shuo power plant, Pingshuo, China), pulverized coal furnace fly ash (PFA) (Datang Taiyuan Second Thermal Power Plant, Tai Yuan, China), activated carbon (AC) (Guang Fu Technology, Tianjin, China), 5A molecular sieve (5A) (Hua Kang, Gongyi, China), and γ-aluminum oxide (γ-Al_2O_3) (Heng Xing, Tianjin, China) were used as supports. The adsorbents were prepared by impregnating. The theoretical loading of K_2CO_3 were selected 10%, 30%, and 50%, respectively. 100 g supports were added to an aqueous solution containing a certain amount of K_2CO_3 (10 g, 30 g, and 50 g) in 500 mL deionized water. Then, it was stirred with a magnetic stirrer (Guang Ming, Beijing, China) for 10 h at room temperature. Thereafter, the mixture was dried in the oven (Gang Yuan, Tianjin, China) for 8 h at 105 °C. The dried samples were then calcined in a muffle furnace (Ke Jing, Zhengzhou, China) for 4 h at 300 °C. Lastly, the samples were crushed and screened to a particle size range within 75 μm by the sieve shaker (Xin Da, Shaoxing, China) to get the final adsorbent.

The actual loading represents the ratio of the active component K_2CO_3 loaded on the support particles in the prepared adsorbent, which is shown in Equation (1). The theoretical and corresponding actual loadings of the five supports are shown in Table 1.

$$L_R = \frac{m_f - m_i}{m_i} \times 100\% \tag{1}$$

where L_R is the actual loading, %; m_f is the mass of the sample after loading, g; mi is the mass of the support particles, g.

Table 1. Actual loading of different supports.

Theoretical Loading (%)	Actual Loading (%)				
	AC	Al$_2$O$_3$	5A	CFA	PFA
10	9.4	7.7	8.4	8.5	8.9
30	28.6	21.2	24.3	26.2	27.5
50	44.3	32.5	37.5	39.6	41.3

In addition, the content of the K element in the adsorbent was analyzed by ARL9800XP X-ray Fluorescence (XRF) (ARL, Berne, Switzerland) and a more accurate actual load was obtained. The results are shown in Table 2. It can be seen from the table that the actual loading calculated by the XRF characterization method is similar to the calculation result of Equation (1) and the results of this paper are verified.

Table 2. Actual loading of different supports by XRF.

Theoretical Loading (%)	Actual Loading (%)				
	AC	Al$_2$O$_3$	5A	CFA	PFA
10	9.63	7.68	8.45	8.64	8.91
30	28.72	21.32	24.24	26.27	27.65
50	44.32	32.45	37.73	39.81	41.36

In order to obtain the microscopic characteristics of the prepared adsorbent, the N_2 adsorption and desorption experiments were carried out by ASAP 2460 analyzer (Micromeritics, Norcross, GA, USA). The specific surface area was calculated by the Brunauer-Emmett-Teller (BET) equation and the pore structure parameters of the adsorbent were obtained by the Barrett-Joyner-Halenda (BJH) method. The surface morphology of the adsorbent was obtained by the Nova Nano SEM 50 scanning electron microscopy (Thermo Fisher Scientific, Hillsboro, OR, USA) and the crystal structure of the adsorbent was obtained by the X/max-2500 X-ray diffractometer (XRD) (Rigaku, Tokyo, Japan).

2.2. Fixed Bed CO$_2$ Adsorption Experiment System

The fixed bed carbon adsorption experimental system is shown in Figure 1. It is mainly composed of a simulated gas production system, a reaction system, and a data acquisition and processing system. The simulated flue gas $N_2/CO_2/H_2O$ in the experiment was provided by a gas distribution system in which the flow rate of water was controlled by a Series III metering pump (SSI, Cincinnati, OH, USA) and gasified into water vapor by electric heating. It was also thoroughly mixed with N_2/CO_2. During the carbonation reaction, the CO_2 concentration was monitored online by the MOT (Monitor) series gas detector (Keernuo, Shenzhen, China).

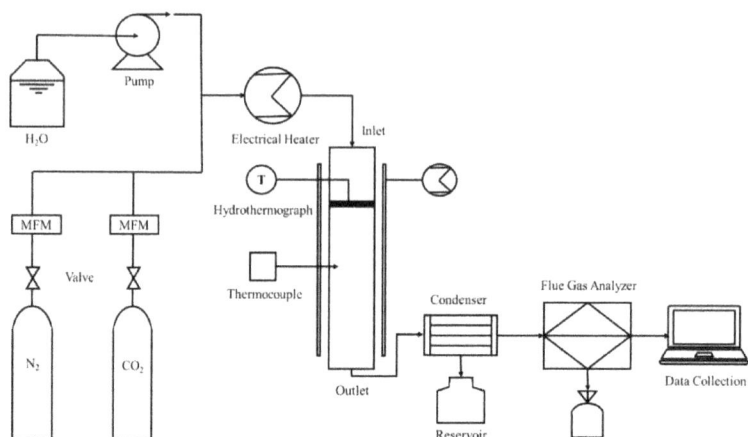

Figure 1. Carbonation reaction system.

In the experiment of carbonation of adsorbent, 5 g adsorbent was placed in the reactor with a simulated flue gas volume of 500 mL (water vapor concentration is 10%). Since the reaction temperature range of carbonation is 50 to 90 °C, considering that the actual flue gas environment of the power plant, the reaction temperatures were selected as 50 °C, 60 °C, 70 °C, 80 °C, and 90 °C while the CO_2 concentrations were selected as 5%, 7.5%, 10%, 12.5%, and 15%, respectively. Furthermore, the CO_2 adsorption characteristics of the adsorbent were evaluated by the CO_2 adsorption rate η as shown in Equation (2).

$$\eta = \frac{\frac{\Delta m}{2 \times M_{KHCO_3}} \times M_{K_2CO_3}}{m_{k_2CO_3}} \times 100\%. \tag{2}$$

where Δm is the weight gain of the sample during the reaction, g. $M_{K_2CO_3}$ is the relative molecular mass of K_2CO_3, $M_r = 138$. M_{KHCO_3} is the relative molecular mass of $HKCO_3$ and $M_r = 100$ $M_{k_2CO_3}$ is the mass of the adsorbent before the reaction, g.

Moreover, in order to obtain the failure characteristics of the adsorbents under different flue gas composition, the adsorbents sample adsorbed with CO_2 were filled into the fixed bed carbon adsorption experiment system as described above to perform multiple adsorption/regeneration cycle experiments. During the cycle experiment, NO (0.05%), SO_2 (0.05%), and HCl (0.05%) were added to the atmosphere of the carbonation reaction. Regeneration experiments were carried out in an N_2 atmosphere and the reaction temperature was 200 °C.

3. Results and Discussion

3.1. Effect of Loading and Adsorption Conditions on Carbonation Reaction

3.1.1. Carbonation Reaction Characteristics of Adsorbents under Different Load Conditions

In order to obtain the optimal loading of adsorbents of different support types, the carbonation reaction of the modified potassium-based adsorbents were carried out at 12.5% CO_2 and 70 °C. The results are shown in Figure 2.

Among all the adsorbents, the K_2CO_3/AC sample has the highest CO_2 adsorption rate under different loading conditions and the actual loading of K_2CO_3 in this adsorbent is also the largest. Among the different adsorbents, when the theoretical loading of K_2CO_3 is 10%, the adsorption rate of CO_2 is in the lower range of 30% to 40%. The CO_2 adsorption rates of the five adsorbents increased significantly as the theoretical loading increased to 30%, which reached 79.8%, 65.3%, 68.2%, 71.9%, and 73.7%, respectively. However, when the theoretical loading reaches 50%, the CO_2 adsorption rate shows a slight downward trend. This is because the loading of the active component on the surface of the support is mainly divided into three stages: unsaturated load, saturated load, and multilayer load. With the increase in K_2CO_3 loading from 10% to 30%, the load of the active component undergoes a process of an unsaturated state to saturation. When the loading increases from 30% to 50%, the load state is switched from a saturated load to a multi-layer load and, at this time, the active component has an overlapping multi-layer load phenomenon on the surface of the support. Hence, when the adsorbent undergoes a carbonation reaction, the outermost active component will first undergo a carbonation reaction to form $KHCO_3$. While $KHCO_3$ is a dense and smooth material and the formed product layer hinders the reaction of CO_2 with the active components of the inner layer, it results in a slight decrease in the adsorption rate of CO_2.

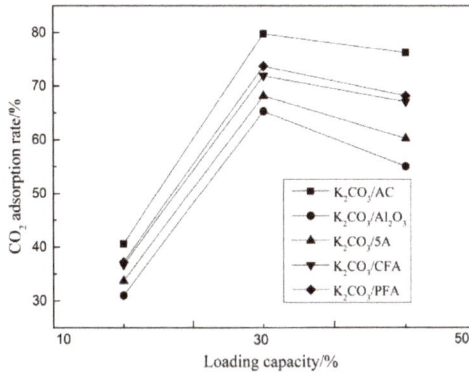

Figure 2. CO_2 adsorption rate at different loading capacities.

3.1.2. Carbonation Reaction Characteristics of Adsorbents under Different Adsorption Conditions

Carbonation reaction experiments of five kinds of adsorbents with the optimal theoretical loading (30%) were carried out under different CO_2 concentrations and reaction temperatures. When studying the effect of different CO_2 concentrations on the adsorption rate, the temperature was set to 70 °C. When studying the effect of different temperatures on the adsorption rate, the CO_2 concentration was set to 12.5% and the results are shown in Figures 3 and 4.

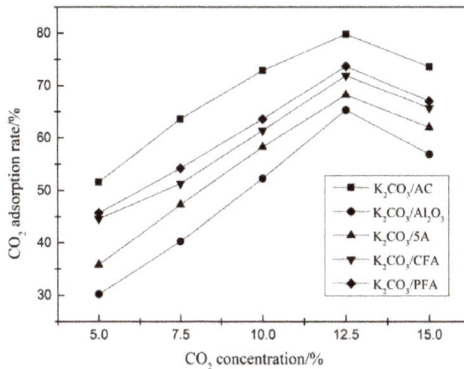

Figure 3. CO_2 adsorption rate at different CO_2 concentrations.

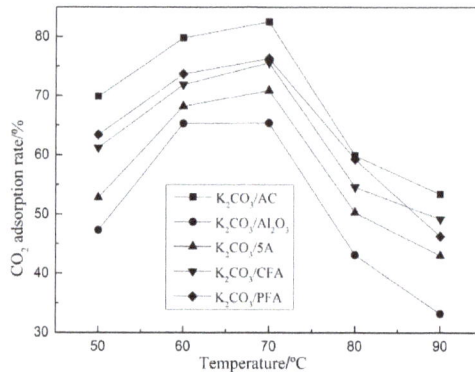

Figure 4. CO_2 adsorption rate at different temperatures.

It can be concluded from Figure 3 that the CO_2 adsorption rate increases first and then decreases with the increase of CO_2 concentration and the optimal CO_2 concentration is 12.5%. This is because the difference between the CO_2 concentration on the surface of the adsorbent and the CO_2 concentration of the atmosphere is a driving force during the entire adsorption process. Hence, increasing the initial concentration of CO_2 can accelerate the adsorption rate, which promotes the CO_2 adsorption performance of the adsorbent. However, when the number and activity of the adsorbent sites on the adsorbents surface are certain, the increase of the concentration of CO_2 will also increase the adsorption capacity of the adsorbents, which leads to the decrease of the adsorption efficiency and the rate of the adsorbents. Therefore, the CO_2 adsorption performance of the adsorbent is inhibited. Among all the adsorbents, the CO_2 adsorption rate of K_2CO_3/AC is the largest and the degree of carbonation reaction of K_2CO_3/Al_2O_3 and $K_2CO_3/5A$ is relatively close while the effect is poor.

It can be seen from Figure 4 that the optimal reaction temperature is 70 °C and the adsorption rates at 60 °C and 70 °C are relatively close. The CO_2 adsorption rate decreases significantly with the increase of temperature at the range of 70 to 90 °C. The adsorption of CO_2 on the supporting potassium-based adsorbent existed in both physical and chemical adsorption and the chemisorption was dominant. With the increase of the reaction temperature, the diffusion process of CO_2 in the surface of the adsorbent and the internal pores is accelerated, which is beneficial to the reaction of CO_2 with the adsorbent. Meanwhile higher temperature leads to easier breakage of chemical bonds and the lower energy barrier of the active component, which promotes chemical reaction. However, the reaction of K_2CO_3, CO_2, and H_2O to form $KHCO_3$ in the range of 120 to 200 °C is reversible. Hence, the activity of K_2CO_3 decreases and the reaction process tends to reverse when the temperature rises from 70 °C to 90 °C, which leads to the rapid decrease of the CO_2 adsorption rate.

In conclusion, the optimal theoretical loading of the adsorbent is 30% and the optimal adsorption reaction conditions include a CO_2 concentration of 12.5% and a reaction temperature of 70 °C. Moreover, among the active adsorbents prepared by five different supports, the K_2CO_3/AC adsorbent has the highest degree of carbonation reaction, which is followed by K_2CO_3/CFA, K_2CO_3/PFA, and $K_2CO_3/5A$ adsorbents. K_2CO_3/Al_2O_3 has the worst activity.

3.2. Microscopic Characteristics

In this paper, the microscopic properties (pore structure and surface morphology) of the adsorbent before and after modification at the optimal theoretical loading were studied. Meanwhile, the composition and crystal structure of the adsorbent before and after carbonation reaction were studied.

3.2.1. Pore Structure

The pore structure parameters affecting the adsorption characteristics of the adsorbent mainly include specific surface area, specific pore volume, and pore size distribution. By taking the adsorbent before and after modification under optimal theoretical loading as the research object. N_2 adsorption/desorption experiment was carried out at a low temperature to study the pore structure of the adsorbent. The results are shown in Table 3. When studying the effect of the pore structure on the ability of the adsorbent to adsorb CO_2, the specific surface area per unit volume Z is introduced to characterize its pore richness [25,26], which is represented by Formula (3) below.

$$Z = \frac{S_0}{V_0} \tag{3}$$

where S_0 is the BET specific surface area of the adsorbent, m^2/g. V_0 is the sum of the specific pore volumes of the adsorbent, cm^3/g.

Furthermore, surface fractal dimension of the adsorbent can be selected as one of the evaluating parameter to characterize its pore structure. When the fractal dimension is 2, the surface of the object is smooth and regular. When the fractal dimension is close to 3, the surface structure becomes disordered and disordered. Its value can be obtained by the FHH (Frenkel, Halsey and Hill) equation [27]. Preifer et al. [28] believe that the FHH theory applies to the adsorption and desorption processes in cryptopores (one to several ten nm). The fractal FHH equation is shown in Equation (4) and the fractal dimension of the internal pore surface of the particle can be determined by the N_2 adsorption isotherm. In the adsorption process, the adsorption interface is mainly affected by the Van der Waals Force in which the relationship between the constant S_N and the fractal dimension D_S is shown in Equation (5). Thus, the fractal dimension can be calculated by simultaneous Equations (4) and (5).

$$\frac{V}{V_m} = k \cdot \left(ln \frac{P_0}{P} \right)^{S_N} \tag{4}$$

$$S_N = D_S - 3 \tag{5}$$

where V_m is the single layer saturated adsorption capacity, k is a constant, P is the pressure of the adsorbate, P_0 is the saturated vapor pressure of the adsorbate, S_N is a constant related to the adsorption mechanism and fractal dimension D, and D_S is the fractal dimension.

From Table 3, the microstructure of K_2CO_3 is poor and its specific surface area and the cumulative pore volume are small. The adsorption of CO_2 by the adsorbent mainly depends on the chemical reaction. In addition, in all the adsorbents after loading, the BET specific surface area and the cumulative pore volume decreased greatly and the pore richness Z also showed a decreasing trend, which indicates that a large amount of K_2CO_3 adhered to the surface and pores of the support after loading.

Since the surface area and cumulative pore volume of AC are much larger than other supports, the active component is most loaded, which is beneficial to the carbonation reaction of the adsorbent after the loading modification. The specific surface area of the two kinds of fly ash decreased from 80.85 m^2/g and 99.70 m^2/g to 47.02 m^2/g and 73.17 m^2/g before and after loading and is only lower than AC and K_2CO_3/AC. However, the cumulative pore volume of both before and after loading is small. When the 5A molecular sieve was used as the support, the specific surface area and cumulative pore volume of the adsorbent before and after loading decreased most clearly, which decreased by 80.94% and 67.88%, respectively. This is because, during the load process, the inside of the molecular sieve is plastically deformed due to surface tension and a large number of active components fill the pores, which results in deterioration of the microstructure. In addition, Al_2O_3 has a smooth and dense structure. Even though the pore structure is improved after loading K_2CO_3, the lifting effect is not significant.

Table 3. Pore structure parameters of adsorbents before and after loading with the optimal theoretical loading.

Samples	BET Specific Surface Area $m^2 \cdot g^{-1}$	Cumulative Pore Volume $cm^3 \cdot g^{-1}$	Fractal Dimension	Pore Richness Z	Relative Pore Volume %	
					Micropore and Mesopore	Macropore
K_2CO_3	0.78	0.0073	2.0987	106.849	99.22	0.78
AC	493.87	0.3738	2.9050	1321.215	95.56	4.44
K_2CO_3/AC-30%	200.94	0.1630	2.9009	1232.761	96.84	3.16
Al_2O_3	18.54	0.0445	2.4512	416.629	93.65	6.35
K_2CO_3/Al_2O_3-30%	7.13	0.0275	2.5325	259.27	98.46	1.54
5A	44.75	0.1012	2.7958	443.061	92.61	7.39
K_2CO_3/5A-30%	8.53	0.0325	2.7789	262.403	96.00	4.00
CFA	80.85	0.0194	2.8775	4167.371	96.31	3.69
K_2CO_3/CFA-30%	47.02	0.0119	2.8543	3950.847	95.44	4.56
PFA	99.70	0.0257	2.8733	3879.494	93.81	6.19
K_2CO_3/PFA-30%	73.17	0.0199	2.8783	3676.935	93.35	6.65

Combined with the results of CO_2 adsorption, it can be concluded that the type of support and the BET specific surface area of the adsorbent after loading modification determine the degree of the carbonation reaction while the cumulative pore volume and pore size distribution have less influence. This is because the process of adsorbing CO_2 by the active component K_2CO_3 on the support is a chemical reaction, which is dominant throughout the adsorption process. The larger the surface area, the more active sites on the surface of the support can adhere to K_2CO_3, which increases the adsorption of CO_2 by the adsorbent. When the cumulative pore volume is large, a large number of active components are attached to the pores. However, $KHCO_3$ formed by the reaction of external K_2CO_3 and CO_2 has a dense and less porous substance, which will hinder the diffusion of CO_2 into the pores and inhibit the continuous carbonation reaction. At the same time, it can be concluded that the fractal dimension and the CO_2 adsorption rate also have a proportional relationship and, with the increase of the fractal dimension, the surface morphology of the adsorbent gradually becomes irregular, which is beneficial for the sufficient contact of the active component with CO_2. This enhances the adsorption of CO_2 by the adsorbent.

3.2.2. Surface Morphology

In this paper, the support and corresponding modified adsorbent samples were observed by SEM and the surface morphology and microstructure of the supports and the adsorbents obtained under different modification conditions were obtained, which is shown in Figures 5–9.

As can be seen from Figure 5, the surface of the support AC has a rough surface and an irregular block structure. After the active component of K_2CO_3 is loaded, the original block structure is broken by the impregnation process and becomes a large number of irregular small particles. The surface and pores of the formed adsorbent are filled with a large amount of active components and the roughness is intensified. However, there are still significant gaps between the particles, which means the overall arrangement is loose.

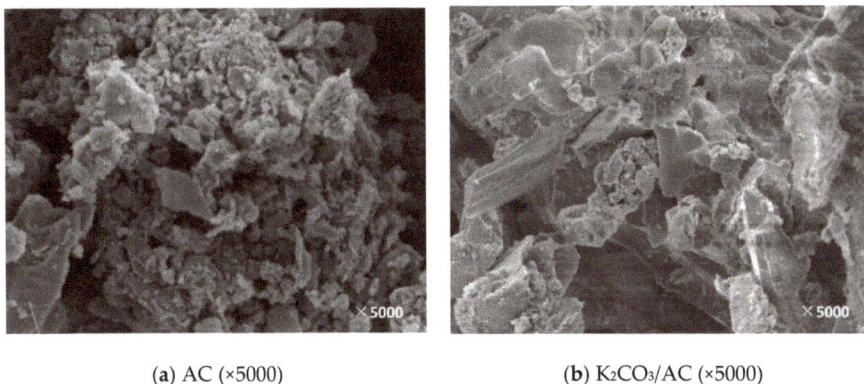

| (a) AC (×5000) | (b) K₂CO₃/AC (×5000) |

Figure 5. SEM image before and after AC loading. (**a**): The SEM image of AC; (**b**): The SEM image of K_2CO_3/AC.

Figure 6 shows the SEM results before and after loading the 5A molecular sieve. The molecular sieve particles before loading have a spherical structure with a rough surface and an abundant pore structure. The pores of the adsorbent after the loading are filled with a large amount of K_2CO_3, which confirmed the results of the previously mentioned reduction in a specific surface area and a cumulative pore volume. At the same time, it can be concluded that K_2CO_3 is loaded in a multi-layered manner in the pores. Therefore, after the surface-active component is completely reacted, the resistance of CO_2 into the internal channel increases and the carbonation reaction is hindered.

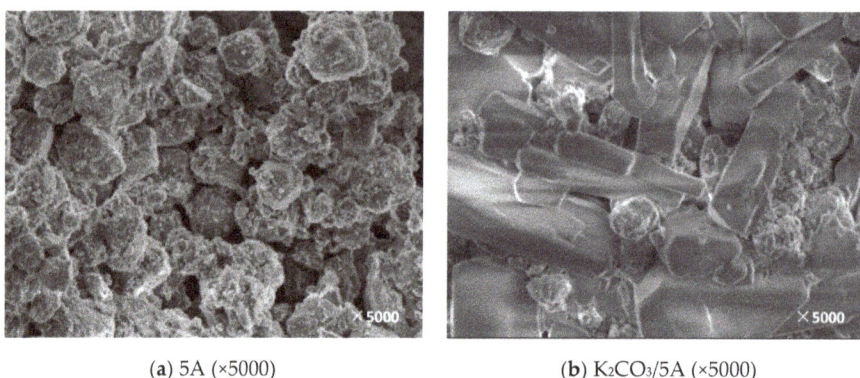

(a) 5A (×5000) (b) K₂CO₃/5A (×5000)

Figure 6. SEM image before and after 5A loading. (**a**): The SEM image of 5A; (**b**): The SEM image of $K_2CO_3/5A$.

Figure 7 shows an SEM image before and after loading Al_2O_3. Compared with other kinds of adsorbents, Al_2O_3 has a dense surface and a relatively regular layered structure. Although the surface structure is slightly improved after the impregnation, it is not significant and only a small amount of the active component is loaded on the surface.

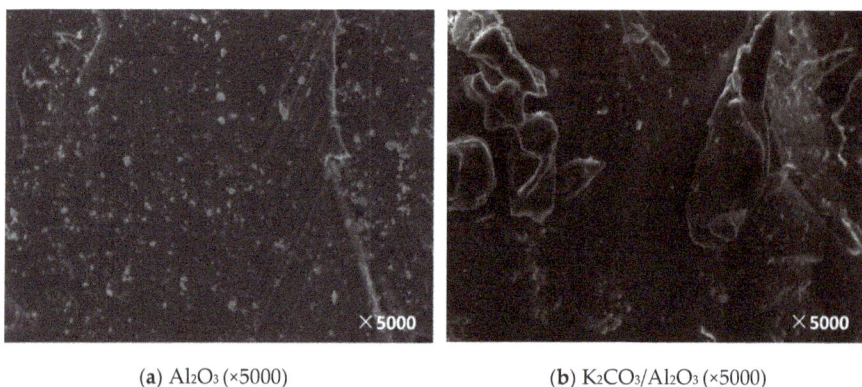

(a) Al₂O₃ (×5000) (b) K₂CO₃/Al₂O₃ (×5000)

Figure 7. SEM image before and after Al_2O_3 loading. (**a**): The SEM image of Al_2O_3; (**b**): The SEM image of K_2CO_3/Al_2O_3.

Figure 8 shows an SEM image before and after the CFA loading. The surface morphology of the fly ash before loading is relatively rough and it is mainly composed of a dense porous structure. After loading, a large amount of the active components are deposited on the surface and the CO_2 can be sufficiently contacted to facilitate the carbonation reaction.

Figure 9 shows the SEM image before and after the PFA loading. The surface of the fly ash before loading contains a large number of round particles. The fly ash particles are fine and the shape is relatively simple. The particle diameter of the particles is different, the maximum is not more than 10 μm, and the diameter of most particles is 1–4 μm [29]. These particles are most likely aluminosilicate or fly ash balls. After the impregnation, the surface morphology changed greatly, the spherical particles disappeared, and the surface was loaded by the active components. However, the overall pore structure did not change significantly.

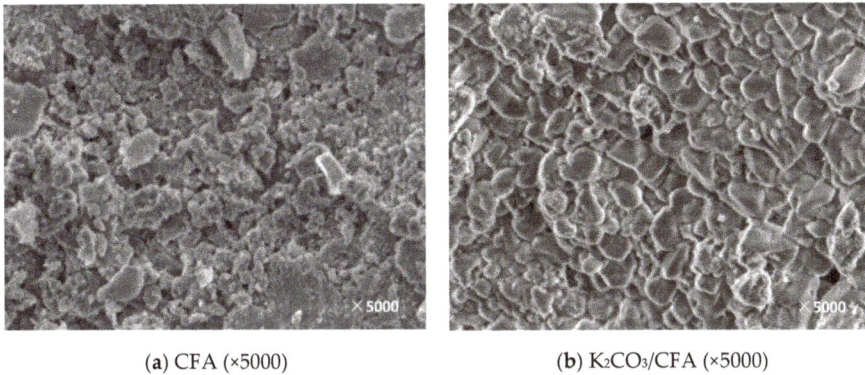

(a) CFA (×5000)

(b) K₂CO₃/CFA (×5000)

Figure 8. SEM image before and after CFA loading. (**a**): The SEM image of CFA; (**b**): The SEM image of K_2CO_3/CFA.

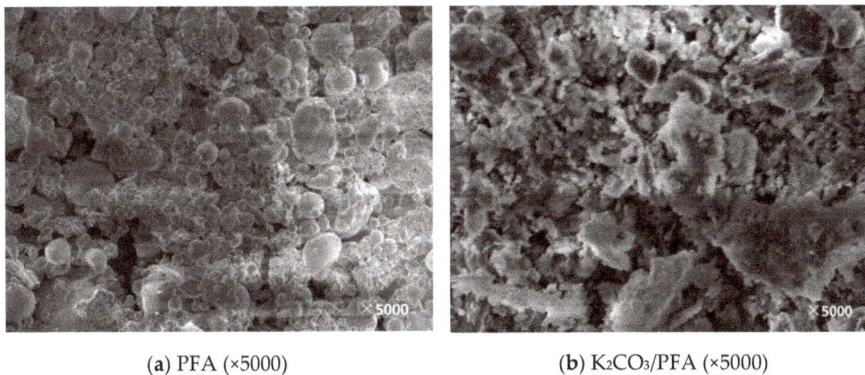

(a) PFA (×5000)

(b) K₂CO₃/PFA (×5000)

Figure 9. SEM image before and after PFA loading. (**a**): The SEM image of PFA; (**b**): The SEM image of K_2CO_3/PFA.

3.2.3. Lattice Structure

In order to obtain the changes of the composition of the K_2CO_3 adsorbent under different carbonation reactions, XRD analysis was performed on the samples before and after adsorption, which is shown in Figure 10. Among them, only the K_2CO_3/Al_2O_3 adsorbent produces a new chemical reaction in the carbonation reaction process and $KAl(CO_3)_2(OH)_2$ is formed. The presence of Al_2O_3 will compete with K_2CO_3 for adsorption. The chemical adsorption of CO_2 by the active component is reduced. The miller index corresponding to the strong diffraction peaks of the material $K_2CO_3 \cdot 1.5H_2O$ before the reaction are (1 1 0), (1 1 3), and (1 1 6), respectively. The miller index corresponding to the three strong peaks of the main product $KHCO_3$ after the reaction are (1 0 4), (2 0 2), and (2 1 1), respectively. The miller index of the K_2CO_3/5A and K_2CO_3/AC adsorbents corresponding to the strong diffraction peaks of $K_2CO_3 \cdot 1.5H_2O$ before the carbonation reaction are (0 2 2), (1 1 2), and (0 2 2), (2 0 0), (1 0 2) respectively. In addition, the miller index corresponding to the strong diffraction characteristic peaks of the main product $KHCO_3$ after the reaction are (1 0 4), (2 0 2), (2 1 1), and (4 0 0), (2 0 1), (−3 1 1). The main components before the CFA reaction are SiO_2 and $CaSO_4$ and the miller index corresponding to the strong diffraction characteristic peaks are (0 0 2) and (1 −1 2), respectively. The main components before the PFA reaction are SiO_2, $Al_6Si_2O_{13}$, and Al_2SiO_5 and the crystal miller index corresponding to the strong diffraction characteristic peaks are (0 0 2), (1 0 2), and (−1 2 1), respectively. The main products of the adsorbents with CFA and PFA as supports are $KHCO_3$

and unreacted K_2CO_3 and the corresponding miller index are (1 3 −1), (0 1 2), and (−3 1 1), (1 1 2), respectively.

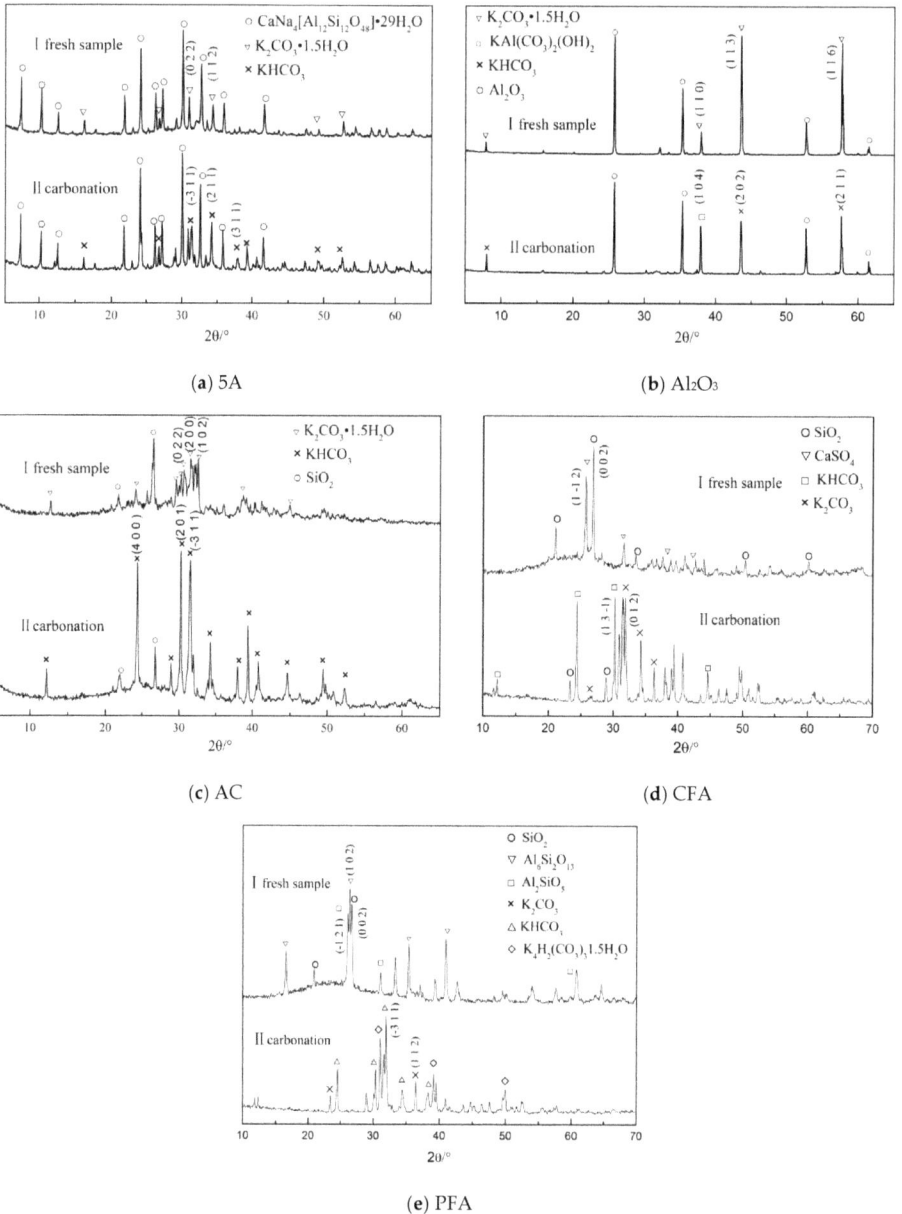

(a) 5A

(b) Al₂O₃

(c) AC

(d) CFA

(e) PFA

Figure 10. XRD diffraction pattern before and after carbonation of different adsorbents. (**a**): The XRD result of 5A; (**b**): The XRD result of Al₂O₃; (**c**): The XRD result of AC; (**d**): The XRD result of CFA; (**e**): The XRD result of PFA.

3.3. Reaction and Adsorption Kinetics

The intrinsic reaction of the carbonation reaction of potassium-based adsorbent is the chemical reaction of K_2CO_3 particles with CO_2 and H_2O. The chemical reaction process of K_2CO_3 absorbing CO_2 is shown in Formula (6). It is a typical gas-solid non-catalytic reaction and the K_2CO_3 particles used in the experiment have a very small specific pore volume and are compact particles. Hence, the shrinking core model was used to describe its carbonation reaction mechanism [30].

$$K_2CO_3(s) + CO_2(g) + H_2O(g) = 2KHCO_3(s) \tag{6}$$

The carbonation reaction of adsorbent mainly includes three basic processes: gas film diffusion, surface adsorption, and intraparticle diffusion. Due to the rapid formation of the product layer with a larger specific volume than K_2CO_3 in the reaction process, the diffusion resistance of the mixed gas at the product layer is much greater than the diffusion resistance of the gas film. Hence, in the research process, the film diffusion process was ignored. Chemical reaction kinetic models and product layer diffusion kinetic models were used to study the reaction mechanism and main control forms. The chemical reaction control process is as shown in Formula (7) and the product layer diffusion control process is shown in Formula (8).

$$t = \frac{\rho_p R_P}{k_s C_A^0}\left[1 - (1 - \eta_C)^{1/3}\right] \tag{7}$$

where k_s is the surface reaction control coefficient, min^{-1}.

$$t = \frac{\rho_p R_P^2}{6D_e C_A^0}\left[1 - 3(1 - \eta_C)^{2/3} + 2(1 - \eta_C)\right] \tag{8}$$

where t is the reaction time and s. ρ_p is the molar density of the absorbent, mol/m^3. R_P is the initial radius of the particle and m. D_e is the diffusion coefficient of the gas in the product layer, m^2/s. C_A^0 is concentration of the reaction gas at time t = 0, mol/m^3. n is the particle conversion rate, %.

In this paper, the shrinking core model was used to fit the carbonation reaction results of five adsorbents under different loading conditions at different reaction temperatures. The results are shown in Figure 11. The error between the relevant parameters obtained by fitting the equation and the experimental values is represented by the correlation coefficient R^2. The larger the value, the closer the description of the adsorption process is to the selected model. The correlation coefficients of all adsorbent samples obtained by fitting are close to 0.99. It can be concluded that the adsorption process of CO_2 on the adsorbent samples at different reaction temperatures is consistent with the nucleation kinetic model.

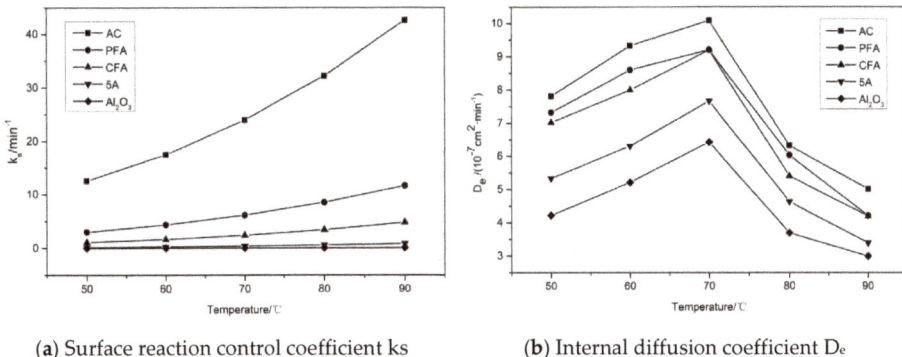

(a) Surface reaction control coefficient ks   (b) Internal diffusion coefficient De

Figure 11. Kinetic parameters of five adsorbents at different temperatures. (a): Surface reaction control coefficient ks; (b): Internal diffusion coefficient De.

From Figure 11, it is concluded that, as the reaction temperature increases, the surface reaction rate constant k_s of the five adsorbents gradually increase. This is because the surface chemical reaction time is prolonged due to an increase in the reaction temperature and the carbonation pellet conversion rate is continuously increased. At the same time, the degree of reaction presents two stages. When the reaction temperature is lower than 70 °C, the diffusion coefficient D_e of the product layer gradually increases. This is because, during the adsorption process, the difference between the concentration of CO_2 and H_2O in the adsorbent surface and the adsorption atmosphere is the driving force in the whole adsorption process and the increase in temperature increases the adsorption rate and promotes the reaction, which accelerates the diffusion process of the reaction gas. In addition, when the reaction temperature is 70 °C, the carbonation conversion rate is increased to the maximum value. When the reaction temperature is higher than 70 °C, D_e shows a decreasing trend. This is because, as the reaction proceeds, the nuclear radius of the unreacted particles of the adsorbent gradually decreases. In addition, since the active material in the adsorbent generates more products during the early carbonation reaction, the formed $KHCO_3$ product layer is more likely to wrap the surface of the adsorbent particles, which hinders the outward diffusion of the reactants in the adsorbent. At the same time, the diffusion of CO_2 and H_2O to the surface of unreacted particles is hindered and the internal diffusion resistance is remarkably enhanced.

In addition, since the carbonation reaction is jointly controlled by diffusion and a chemical reaction, the main control form of the reaction model is determined by the ratio $\rho(\rho = (1/A_2)/(1/A_1) = (\frac{\rho_p R_P}{k_s C_A^0})/(\frac{\rho_p R_P^2}{6 D_e C_A^0}) = \frac{6 D_e}{k_s R_P})$ of the obtained chemical reaction rate constant A_1 and the diffusion rate constant A_2. When $\rho \ll 1$, the reaction is mainly controlled by the intrinsic chemical reaction. When $\rho \gg 10$, the external mass diffusion is ignored and the reaction is controlled by the diffusion process through the product layer. When ρ is between 1 and 10, the reaction is not only controlled by the intrinsic chemical reaction but also by the diffusion process through the product layer. The reaction is jointly controlled by both [31]. When the reaction temperatures are 50 °C, 60 °C, and 70 °C, respectively, the ρ values are 0.3569, 0.4623, and 0.8651, respectively, which indicates that the reaction is mainly controlled by the surface chemical reaction when the reaction temperature is lower than 70 °C. When the temperatures are 80 °C and 90 °C, the ρ values are 1.2963 and 1.3387, respectively, and the reaction is controlled by two processes, which verifies the above results.

In addition, in order to study the effect of temperature on the carbonation reaction process of the adsorbent, the Arrhenius equation was used. In addition, in the process of studying the influence of temperature on the carbonation reaction of adsorbent, the activation energy was obtained by the Arrhenius equation, which is shown in Formula (9). During the calculation, the logarithm of the two sides of the Arrhenius equation is obtained to obtain Equation (10). By using $1/T$ as the abscissa and $\ln(ks)$ as the ordinate for linear fitting, the adsorption activation energy during the adsorption reaction can be obtained. The corresponding fitting results are shown in Figure 12.

$$k_s = A_{r1} \exp(-E_{a1}/RT) \tag{9}$$

where T is the reaction temperature, K. A_{r1} is the pre-exponential factor of the surface reaction process, $\min^{-1}$. E_{a1} is the activation energy of the surface reaction process, kJ/mol. R is the gas constant, 8.314 J/(mol·K).

$$\ln(k_s) = -E_{a1}/RT + C \tag{10}$$

where C is a constant.

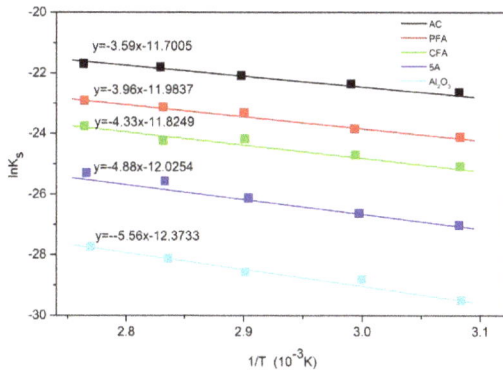

Figure 12. Arrhenius relational fitting diagram for a chemical reaction control phase.

The activation energies of K_2CO_3/AC, K_2CO_3/PFA, K_2CO_3/CFA, $K_2CO_3/5A$, and K_2CO_3/Al_2O_3 are 29.82 kJ/mol, 32.91 kJ/mol, 36.02 kJ/mol, 40.57 kJ/mol, and 46.21 kJ/mol, respectively. Activation energy refers to the energy required for a molecule to change from a normal state to an active state in which a chemical reaction is likely to occur, which reflects the difficult degree of a chemical reaction. Among the five adsorbents, K_2CO_3/AC has the lowest activation energy and the carbonation reaction is the easiest to carry out, so the reaction rate and carbonation degree are the largest. Similarly, the activation energy of the K_2CO_3/Al_2O_3 adsorbent is the highest and the carbonation reaction is difficult, which results in the lowest degree of carbonation of the adsorbent.

It can be seen from the above information that the adsorption process of CO_2 by adsorbents mainly includes three basic processes: external mass transfer, surface adsorption, and intraparticle diffusion. Hence, the pseudo-first order kinetic model, the pseudo-second order kinetic model, the intra-particle diffusion model, and the Elovich model are used to study the decarburization mechanism of different kinds of adsorbents and determine the rate-determining step in the adsorption process. The pseudo-first order kinetic model and the intra-particle diffusion model mainly study the physical adsorption process while the pseudo-second order kinetic model and the Elovich model are mainly used to study chemisorption. Among them, pseudo-first order kinetics mainly studies the external mass transfer process, which is shown in Formula (11). The pseudo-second order kinetic model is based on the Langmuir adsorption isotherm equation to study the formation of chemical bonds to verify that the adsorption process is dominated by chemisorption, which is shown in Formula (12). The intra-particle diffusion model is derived from the mass balance equation, which mainly studies the internal diffusion process of the pores during solid adsorption, as shown in Equation (13). The Elovich model is based on the Temkin adsorption isotherm equation and mainly describes the chemisorption process. It is similar to the pseudo-second order and the fitting results of the two models can be used to verify the accuracy of each other, which is shown in Equation (14).

$$q = q_e\left(1 - e^{-tk_1}\right) \tag{11}$$

where q is the adsorption amount of the adsorbent per unit mass at time t, g/g. q_e is the adsorption amount of the adsorbent per unit mass in equilibrium, g/g. t is the adsorption time, min. k_1 is the pseudo-first order rate constant, min^{-1}.

$$q = \left(q_e^2 k_2 t\right)/(1 + q_e k_2 t) \tag{12}$$

where k_2 is the pseudo-second order rate constant, ng/(g·min).

$$q = k_{id}t^{1/2} + C \tag{13}$$

where k_{id} is the intra-particle diffusion rate constant, $g/(g \cdot min^{1/2})$. C is the constant related to the thickness of the boundary layer, g/g. It decreases with the increase of heterogeneity and hydrophilic groups of the adsorbent surface. The larger the value, the greater the influence of the boundary layer on the adsorption.

$$q = (1/\beta)ln(t + t_0) - (1/\beta)ln(t_0) \tag{14}$$

where α is the initial adsorption rate, $g/(g \cdot min^{1/2})$. β is a constant related to the surface coverage and activation energy, ng/g. $t_0 = 1/(\alpha \cdot \beta)$.

The four adsorption kinetic models were used to calculate and fit the adsorbent carbon adsorption experimental data. The results are shown in Table 4. The correlation coefficients R_2 of different adsorbents are all close to 0.99. It can be concluded that the adsorption process of CO_2 by different kinds of adsorbents is consistent with these four kinetic models. The adsorption process is affected by both physical adsorption and chemical adsorption. The adsorption of CO_2 is not a single monolayer layer adsorption, but it is related to the adsorption site of the adsorbent. Moreover, through the predicted equilibrium adsorption amount q_e in the pseudo-first order and pseudo-second order kinetic models, it can be concluded that the adsorption process of CO_2 by the five adsorbents do not reach a saturation state within 60 min and q_e has a positive correlation with its actual adsorption amount, which verifies the correctness of the fitting results. The fitting coefficient of the pseudo-first order kinetic model of K_2CO_3/Al_2O_3-30% sample is higher than that of the pseudo-second order kinetic model, which indicates that the rate-determining step is mainly a physical adsorption process, but its pseudo-first order and pseudo-second order rate constants are low, as mentioned above. This is mainly due to the poor surface pore structure and the low content of active substances. In addition, the rate-determining step of the other adsorbents is mainly chemical adsorption.

Table 4. Fitting parameters of different sorbents.

Sorbents (30% Loading)	Pseudo-First Order Kinetic Equation			Pseudo-Second Order Kinetic Equation			Intra-Particle Diffusion Kinetic Equation			Elovich Equation		
	R^2	k_1	q_e	R^2	k_2	q_e	R^2	k_{id}	c	R^2	α	β
K_2CO_3/5A	0.9931	2.38×10^{-4}	1379	0.9985	2.25×10^{-9}	1793	0.9728	12.0765	−26	0.9973	0.3901	2.01×10^{-3}
K_2CO_3/Al_2O_3	0.9975	3.31×10^{-5}	1138	0.9956	1.75×10^{-9}	1367	0.8789	8.4099	−208	0.9712	0.3784	3.15×10^{-3}
K_2CO_3/CFA	0.9994	6.28×10^{-4}	2429	0.9999	2.59×10^{-7}	3605	0.9969	20.1509	−294	0.9995	0.5409	3.31×10^{-4}
K_2CO_3/PFA	0.9998	7.09×10^{-4}	5524	0.9999	5.12×10^{-7}	8386	0.9962	32.3145	−727	0.9993	0.5851	7.92×10^{-4}
K_2CO_3/AC	0.9995	7.39×10^{-4}	6404	0.9999	6.46×10^{-7}	10199	0.9948	36.4391	−957	0.9991	0.8879	2.54×10^{-4}

In addition, the intra-particle diffusion model was used to fit the cumulative CO_2 adsorption amount per unit mass of different adsorbents and the results are shown in Figure 13. With the increase of adsorption time, the overall trend of k_{id} is increasing and the actual adsorption rate of CO_2 decreases with the increase of adsorption time, the contradiction between the CO_2 adsorption rate, and the internal diffusion rate indicates that there is surface adsorption during CO_2 adsorption. Hence, the CO_2 adsorption process can be divided into two stages: surface adsorption stage and internal diffusion adsorption stage. In the initial adsorption stage, the surface adsorption is the main form of adsorption because a large number of adsorption active sites exist on the surface of the adsorbent. Therefore, the surface adsorption rate is faster while the internal diffusion rate is smaller, which indicates that, in this stage, intra-particle diffusion does not play a leading role. When the active sites of the surface are occupied, the second stage of adsorption is carried out and diffusion adsorption occurs in the pore. At this time, the micropores and mesopores provide the adsorption active sites of CO_2. Therefore, the adsorption rate is continuously decreased and the internal diffusion rate is increased. Moreover, the fitting curves of all intra-particle diffusion models have not passed through the origin, which is quite different from the experimental results. This indicates that the internal diffusion model cannot describe the adsorption process of CO_2 on the adsorbent surface. The internal diffusion process is not the rate-determining step. The correlation coefficients obtained by the internal diffusion model fitting are small and significantly lower than the correlation coefficients obtained by fitting the pseudo-first order kinetic model, which indicates that the external mass transfer process is the rate-determining step for the adsorption of CO_2 on the surface of the adsorbent relative to the internal diffusion process. In addition, although the fitting curves of the pseudo-second order model can be well matched with the experimental results, the correlation coefficients is slightly lower than the correlation coefficients obtained by the pseudo-first order model. In addition, the equilibrium CO_2 adsorption amount obtained by fitting the pseudo-first order kinetic model is closer to the experimental results. Hence, the conclusion that the external mass transfer is the rate-determining step of CO_2 adsorption on the adsorbent surface is further verified and the adsorption of CO_2 at the active sites also plays a more important role. In addition, the fitting curves of the Elovich dynamic model are also in good agreement with the experimental results, which verifies the existence of the chemisorption process at the active site. The Elovich equation is based on the Temkin adsorption isotherm equation. Hence, it can be considered that the adsorption of CO_2 on the adsorbent surface also follows the Temkin adsorption isotherm equation.

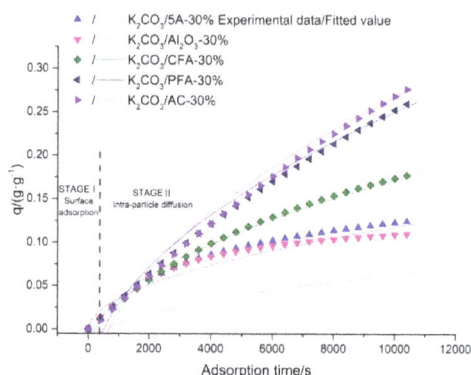

Figure 13. Fitting of intra-particle diffusion kinetic equation.

3.4. Study on the Effect of Flue Gas Composition on the Failure Characteristics of Adsorbents

In this paper, the adsorption/regeneration cycles of K_2CO_3/AC, K_2CO_3/Al_2O_3, $K_2CO_3/5A$, K_2CO_3/CFA, and K_2CO_3/PFA adsorbents were carried out and the degree of failure was characterized by the failure rate η_F, which is shown in Formula (15). In addition, in practical application research, it is found that when $\eta_F > 20\%$, it indicates that the adsorbent has insufficient ability to capture CO_2. Hence, in the study of failure characteristics in this paper, 20% is selected as the critical value of adsorbent failure.

$$\eta_F = \left(1 - \frac{\eta_n}{\eta_0}\right) \times 100\% \tag{15}$$

In the formula, η_n is the adsorption rate of CO_2 after n times of cycle, %. η_0 is the adsorption rate of CO_2 in the first carbonation reaction, %.

Figure 14 shows the variation of the failure rate of different adsorbents with the number of cycles. It can be concluded that the failure characteristics of the five adsorbents are basically the same. The failure rate of all adsorbents in the first 10 cycles increases slowly, but, as the number of cycle increases, the failure rate increases. The K_2CO_3/AC adsorbent has the highest number of cycles when the failure rate reaches more than 20%. The failure rate was only 21.8% until the 23rd cycle. In contrast, K_2CO_3/Al_2O_3 has the worst cycle failure characteristic and the failure rate after 14 cycles is as high as 21.2%.

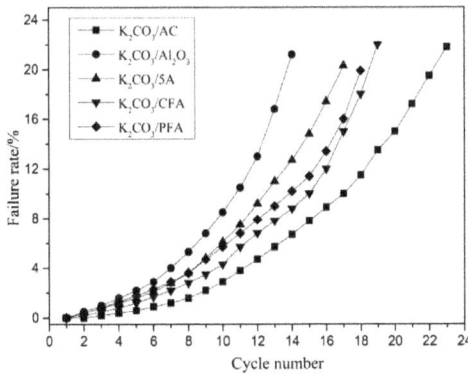

Figure 14. Cyclic experimental results of different adsorbents.

In addition, in the study of the influence of flue gas composition on the failure characteristics of adsorbents, it is found that, in the presence of SO_2, K_2CO_3 and SO_2 will react similar to that in Formula (16) and Formula (17). HCl also reacts with K_2CO_3 to render the adsorbent failure, as shown in Formula (18). NO is stable at low temperatures and does not participate in the reaction with the active components. The five types of supports cannot react with the three acid gases under low temperature and normal pressure. Therefore, the main factors causing the difference in the failure rate are the adsorbent loading and its own CO_2 adsorption rate.

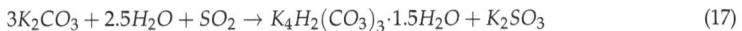

$$K_2CO_3 + SO_2 \rightarrow K_2SO_3 + CO_2 \tag{16}$$

$$3K_2CO_3 + 2.5H_2O + SO_2 \rightarrow K_4H_2(CO_3)_3 \cdot 1.5H_2O + K_2SO_3 \tag{17}$$

$$K_2CO_3 + 2HCl \rightarrow 2KCl + H_2CO_3 \tag{18}$$

4. Conclusions

(1) The actual loading of K_2CO_3/AC adsorbent is the largest and the adsorption performance of CO_2 is the best, which is followed by the K_2CO_3/PFA, the K_2CO_3/CFA, and the $K_2CO_3/5A$ adsorbent and the load modification of K_2CO_3/Al_2O_3 adsorbent is poor while the CO_2 adsorption rate is the lowest.

(2) As the theoretical loading of the five adsorbents increases from 10% to 50%, the CO_2 adsorption rate first increases and then decreases. The best theoretical loading is 30% and the reaction temperature of 70 °C and the concentration of 12.5% CO_2 are the best reaction conditions.

(3) The microstructure of the adsorbents are different after modification of different supports and the cumulative pore volume plays a more important role in the adsorption process than the specific surface area. Among them, K_2CO_3 reacts with Al_2O_3 to produce $KAl(CO_3)_2(OH)_2$ and the support and active component compete for adsorption, which reduces the chemisorption of CO_2.

(4) With the increase of the reaction temperature, the internal diffusion resistance is significantly enhanced. K_2CO_3/AC has the lowest activation energy and the carbonation reaction is the easiest. The adsorption process of CO_2 by the adsorbent is affected by both physical adsorption and chemical adsorption and CO_2 adsorption is related to the adsorption sites of the adsorbent rather than a single monolayer adsorption.

(5) In the simulated flue gas, SO_2 and HCl react with K_2CO_3 to produce new substances, which causes the adsorbent to gradually fail and K_2CO_3/AC has the best cycle failure performance.

Author Contributions: Conceptualization, B.-g.F. Funding acquisition, B.-g.F. Investigation, L.J., Y.-l.W., R.Z. and Y.J. Project administration, L.J. and Y.J. Resources, X.-s.M. and Y.-y.L. Writing—review & editing, L.J.

Funding: The authors acknowledge the financial support for this work provided by the National Natural Science Foundation of China (No. U1510129), the National Natural Science Foundation of China (No. U1510135), the Science and technology major projects of Shanxi Province (No. MD2015-04).

References

1. Le Quéré, C.; Andrew, R.M.; Canadell, J.G.; Sitch, S.; Korsbakken, J.I.; Peters, G.P.; Manning, A.C.; Boden, T.A.; Tans, P.P.; Houghton, R.A.; et al. Global Carbon Budget 2016. *Earth Syst. Sci. Data* **2016**, *8*, 605–649. [CrossRef]
2. Pardemann, R.; Meyer, B. Pre-Combustion Carbon Capture. In *Handbook of Clean Energy Systems*; John Wiley & Sons, Ltd.: Hoboken, NJ, USA, 2015.
3. Hu, J.; Galvita, V.V.; Poelman, H.; Marin, G.B. Advanced chemical looping materials for CO_2 utilization: A review. *Materials* **2018**, *11*, 1187. [CrossRef] [PubMed]
4. Thiruvenkatachari, R.; Su, S.; An, H.; Yu, X.X. Post combustion CO_2 capture by carbon fibre monolithic adsorbents. *Prog. Energy Combust. Sci.* **2009**, *35*, 438–455. [CrossRef]
5. Bhown, A.S.; Freeman, B.C. Analysis and status of post-combustion carbon dioxide capture technologies. *Environ. Sci. Technol.* **2011**, *45*, 8624–8632. [CrossRef] [PubMed]
6. Bu, C.; Gómez-Barea, A.; Leckner, B.; Chen, X.; Pallarès, D.; Liu, D.; Lu, P. Oxy-fuel conversion of sub-bituminous coal particles in fluidized bed and pulverized combustors. *Proc. Combust. Inst.* **2017**, *36*, 3331–3339. [CrossRef]
7. Buhre, B.J.; Elliott, L.K.; Sheng, C.D.; Gupta, R.P.; Wall, T.F. Oxy-fuel combustion technology for coal-fired power generation. *Prog. Energy Combust. Sci.* **2005**, *31*, 283–307. [CrossRef]
8. Cao, Y.; Zhang, H.; Song, F.; Huang, T.; Ji, J.; Zhong, Q.; Chu, W.; Xu, Q. UiO-66-NH2/GO composite: Synthesis, characterization and CO_2 adsorption performance. *Materials* **2018**, *11*, 589. [CrossRef] [PubMed]
9. Cai, Y.; Wang, W.; Li, L.; Wang, Z.; Wang, S.; Ding, H.; Zhang, Z.; Sun, L.; Wang, W. Effective capture of carbon dioxide using hydrated sodium carbonate powders. *Materials* **2018**, *11*, 183. [CrossRef] [PubMed]
10. Sun, R.; Li, Y.; Liu, H.; Wu, S.; Lu, C. CO_2 capture performance of calcium-based sorbent doped with manganese salts during calcium looping cycle. *Appl. Energy* **2012**, *89*, 368–373. [CrossRef]
11. Chen, H.; Zhang, P.; Duan, Y.; Zhao, C. CO_2 capture of calcium based sorbents developed by sol–gel technique in the presence of steam. *Chem. Eng. J.* **2016**, *295*, 218–226. [CrossRef]
12. Sánchez-Zambrano, K.S.; Lima Duarte, L.; Soares Maia, D.A.; Vilarrasa-García, E.; Bastos-Neto, M.; Rodríguez-Castellón, E.; Silva de Azevedo, D.C. CO_2 capture with mesoporous silicas modified with amines by double functionalization: Assessment of adsorption/desorption cycles. *Materials* **2018**, *11*, 887. [CrossRef] [PubMed]
13. Chiang, Y.-C.; Chen, Y.-J.; Wu, C.-Y. Effect of relative humidity on adsorption breakthrough of CO_2 on activated carbon fibers. *Materials* **2017**, *10*, 1296. [CrossRef] [PubMed]

14. Liang, Y.; Harrison, D.P.; Gupta, R.P.; Green, D.A.; McMichael, W.J. Carbon dioxide capture using dry sodium-based sorbents. *Energy Fuels* **2004**, *18*, 569–575. [CrossRef]

15. Querejeta, N.; Plaza, M.G.; Rubiera, F.; Pevida, C. Water vapor adsorption on biomass based carbons under post-combustion CO_2 capture conditions: Effect of post-treatment. *Materials* **2016**, *9*, 359. [CrossRef] [PubMed]

16. Ryu, C.K.; Lee, J.B.; Eom, T.H.; Baek, J.I.; Eom, H.M.; Yi, C.K. CO_2 capture from flue gas using dry regenerable sorbents. In Proceedings of the 8th International Conference on Greenhouse Gas Control Technology, Trondheim, Norway, 19–22 June 2006.

17. Zhao, C.; Chen, X.; Zhao, C. Characteristics of regeneration reaction of dry potassium-based sorbent for CO_2 capture. *J. Eng. Thermophys.* **2009**, *30*, 2145–2148.

18. Zhou, L.; Fan, J.; Shang, X. CO_2 Capture and separation properties in the ionic liquid 1-n-Butyl-3-methylimidazolium nonafluorobutylsulfonate. *Materials* **2014**, *7*, 3867–3880. [CrossRef] [PubMed]

19. Zhao, C.; Chen, X.; Zhao, C. Carbonation reaction characteristics of dry potassium-based sorbent for CO_2 capture. *J. Chem. Ind. Eng.* **2008**, *59*, 2328–2333. [CrossRef]

20. Zhao, C.; Chen, X.; Anthony, E.J.; Jiang, X.; Duan, L.; Wu, Y.; Dong, W.; Zhao, C. Capturing CO_2 in flue gas from fossil fuel-fired power plants using dry regenerable alkali metal-based sorbent. *Prog. Energy Combust. Sci.* **2013**, *39*, 515–534. [CrossRef]

21. Li, Y.; Sun, N.; Li, L.; Zhao, N.; Xiao, F.; Wei, W.; Sun, Y.; Huang, W. Grafting of amines on ethanol-extracted SBA-15 for CO_2 adsorption. *Materials* **2013**, *6*, 981–999. [CrossRef] [PubMed]

22. Zhao, C.; Guo, Y.; Li, C.; Lu, S. Removal of low concentration CO_2, at ambient temperature using several potassium-based sorbents. *Appl. Energy* **2014**, *124*, 241–247. [CrossRef]

23. Lee, S.C.; Choi, B.Y.; Lee, S.J.; Jung, S.Y.; Ryu, C.K.; Kim, J.C. CO_2 absorption and regeneration using Na and K based sorbents. *Stud. Surf. Sci. Catal.* **2004**, *153*, 527–530. [CrossRef]

24. Zhao, C.; Chen, X.; Zhao, C.; Wu, Y.; Dong, W. K_2CO_3/Al_2O_3 for Capturing CO_2 in Flue Gas from Power Plants. Part 3: CO_2 Capture Behaviors of K_2CO_3/Al_2O_3 in a Bubbling Fluidized-Bed Reactor. *Energy Fuels* **2012**, *26*, 3062–3068. [CrossRef]

25. Fan, B.G.; Jia, L.; Li, B.; Yao, Y.X.; Huo, R.P.; Zhao, R.; Qiao, X.L.; Jin, Y. Study on the effects of the pyrolysis atmosphere on the elemental mercury adsorption characteristics and mechanism of biomass char. *Energy Fuels* **2018**, *32*, 6869–6878. [CrossRef]

26. Jia, L.; Fan, B.G.; Li, B.; Yao, Y.X.; Huo, R.P.; Zhao, R.; Qiao, X.L.; Jin, Y. Effects of pyrolysis mode and particle size on the microscopic characteristics and mercury adsorption characteristics of biomass Char. *Bioresources* **2018**, *13*, 5450–5471.

27. Alghamdi, A.A.; Alshahrani, A.F.; Khdary, N.H.; Alharthi, F.A.; Alattas, H.A.; Adil, S.F. Enhanced CO_2 Adsorption by Nitrogen-Doped Graphene Oxide Sheets (N-GOs) prepared by employing polymeric precursors. *Materials* **2018**, *11*, 578. [CrossRef] [PubMed]

28. Pfeifer, P. Fractals in Surface Science: Scattering and Thermodynamics of Adsorbed Films. In *Chemistry and Physics of Solid Surfaces VII*; Springer: Berlin/Heidelberg, Germany, 1988; pp. 283–305.

29. Ji, C.; Huang, X.; Li, L.; Xiao, F.; Zhao, N.; Wei, W. Pentaethylenehexamine-loaded hierarchically porous silica for CO_2 adsorption. *Materials* **2016**, *9*, 835. [CrossRef] [PubMed]

30. Fan, B.G.; Jia, L.; Li, B.; Huo, R.P.; Yao, Y.X.; Han, F.; Qiao, X.L.; Jin, Y. Study on desulfurization performances of magnesium slag with different hydration modification. *J. Mater. Cycles Waste Manag.* **2018**, *20*, 1771–1780. [CrossRef]

31. Jia, L.; Fan, B.G.; Huo, R.P.; Li, B.; Yao, Y.X.; Han, F.; Qiao, X.L.; Jin, Y. Study on quenching hydration reaction kinetics and desulfurization characteristics of magnesium slag. *J. Clean. Prod.* **2018**, *190*, 12–23. [CrossRef]

materials

MDPI

Article

Dust Loading Performance of a Novel Submicro-Fiber Composite Filter Medium for Engine

Jin Long [1,2], Min Tang [1,2,*], Zhaoxia Sun [1,2], Yun Liang [1,2] and Jian Hu [1,2]

[1] State Key Laboratory of Pulp and Paper Engineering, South China University of Technology, Guangzhou 510641, China; lj_king@139.com (J.L.); careysun988@163.com (Z.S.); liangyun@scut.edu.cn (Y.L.); ppjhu@scut.edu.cn (J.H.)

[2] Filtration and Wet Nonwoven Composite Materials Engineering Research Center of Guangdong Province, Guangzhou 510641, China

* Correspondence: tangminde@163.com; Tel.: +86-13763392757

Received: 18 September 2018; Accepted: 15 October 2018; Published: 19 October 2018

Abstract: Airborne dust can cause engine wear and contribute to engine gas emission. This study developed a novel submicro-fiber filter medium to provide protection to engines against dust. The wet-laid submicro-fiber medium was prepared by a dual-layer paper machine, and its dust loading performance was compared with other filter media during laboratory and field tests. During the laboratory tests, the dust holding capacity of the wet-laid submicro-fiber medium was 48% and 10% higher than that of the standard heavy-duty medium and electrospun submicro-fiber medium, respectively. During the field tests, the pressure drop of the wet-laid submicro-fiber filter was 45% lower than that of the standard heavy-duty filter after 10,000 km of operation. It was found that there were two crucial ways to design a better filter medium for protection against dust. Firstly, the surface loading rather than the depth loading was preferred for dust filtration. The submicro-fiber layer kept large amounts of dust particles from penetrating into the depth of filter medium. Secondly, particles were captured preferably by fibers rather than pores. The unique fibrous structure of the wet-laid submicro-fiber medium made more particle deposition take place on fibers via interception and inertial impaction.

Keywords: submicro-fiber; airborne dust; engine filtration; loading performance

---

1. Introduction

Arid and semi-arid areas, which cover approximately one-third of land surfaces, are a major source of mineral dust [1]. If meteorological conditions are favorable, the dust is mobilized and further emitted into the atmosphere. Besides natural processes, a significant amount of dust contaminants are generated by anthropogenic activities, such as farming, industrial manufacturing and mining operations [2]. For motor vehicles, dust suspended in the air is hazardous and can degrade vehicle performance because the combustion engine of the vehicle is exposed to large quantities of air. Airborne dust (quartz) is highly abrasive and is the most common cause of high wear of critical parts, such as piston rings and cylinder walls in engines [3]. Studies have shown that engine life is dependent upon the cleanliness of the air taken in [4]. Engine wear is produced by particles in the size range of 1–40 μm, where the most harmful particles are in the range of 1–20 μm [5]. Dust particles may also significantly contribute to the total engine emissions, including crankcase emissions [6]. According to Schilling [7], 30% of contaminants entering the engine can pass out of the exhaust.

Air filtration is applied to provide protection against the effects of contaminated air. The effects of the engine intake air filter can be both positive and negative [8]. The positive side is to prevent the decrease in engine performance, however, placing an air filter in the path of air coming into the engine can cause a pressure drop, leading to extra energy costs and reduced engine performance.

The pressure drop is proportional to the energy consumption of the air filter, according to Eurovent 4/21 (2014) [9]. The challenge is to achieve a lower pressure drop and longer filter lifetime while removing a satisfactory number of particles.

In the combustion engine industry, fibrous filter media are used to prevent airborne particles from entering the engine. Cellulose-type filter media, made of relatively large fibers with fiber diameters, usually larger than 10 microns, are often used in engine air filtration. To improve filtration performance, the application of ultrafine fibers has become of interest in academic research and within the industry [10–13]. Many theoretical calculations and experimental results of submicro-fiber filter medium have shown that the filtration efficiency could be dramatically improved by the gas slip effect and a large specific surface area [14]. Recently, nanofiber sponge or aerogel were introduced as new porous materials for filtration [15,16]. A popular method of producing submicro-fiber composite filter medium is adding a submicro-fiber layer via electrospinning [17–20]. Donaldson Company (Bloomington, MN, USA) had successfully introduced electrospinning submicro-fiber technology in industrial applications in 1981. However, there are still many challenges, such as high cost, submicro-fiber layer adhesion, and uniformity, and hazards associated with solvent removal and disposal [6]. The mechanical strength of the thin electrospun submicro-fiber layer is poor and can be damaged during the transportation process [21].

The advantage of using a submicro-fiber provides a valid case for relatively clean filters. When dust deposits on submicro-fiber, the benefit of a low pressure drop diminishes with increasing amounts of deposited dust. For military vehicles in high dust concentration environments, a self-cleaning method can be used to blow off dust cake using a pulsejet and make the filter medium clean again [22]. However, most civilian vehicles are not equipped with self-cleaning devices, which means that the air filter would capture dust particles constantly without dust cake removal. There has been a great deal of effort towardsperformance evaluation of a clean submicro-fiber composite without aerosol loading, including submicro-fiber/microfiber mixture [23] and various stack-up nanocomposite [24,25]. Leung et al. [26] studied microfiber/submicro-fiber composites under nano-aerosol loading. However, very limited study has been conducted on submicro-fiber composites against coarse particles that are more harmful to the engine. It is important to design a submicro-fiber medium with a slowly increasing rate of pressure drop in the presence of a high concentration of dust.

To develop a new submicro-fiber medium, this study will combine the ideas of submicro-fiber and multilayer papermaking to laminate a submicro-fiber layer on the substrate via the wet-laid process. During the wet-laid process, resin impregnation for a submicro-fiber layer can be easily applied, which will result in a robust submicro-fiber layer and strong adhesion with a substrate. Langner and Greiner [27] discussed the application of electrospun ultrafine fiber in wet-laid process for filtration. The wet-laid process is also able to use submicro-fibers with superior filtration performance, such as glass wool fiber, which is the main component of cleanroom high efficiency particulate air filters [28]. This paper will present a new method to prepare composite filter medium with unique submicro-fiber layer via paper machine. The preparation of a submicro-fiber composite medium was conducted on an industrial paper machine. The loading performance and dust deposition behavior of the novel submicro-fiber medium were compared with a commercial standard heavy-duty medium and an electrospun submicro-fiber medium to provide a fundamental discussion on the submicro-fiber structure design against high dust concentrations.

2. Materials and Methods

2.1. Preparation of Wet-laid Submicro-Fiber Composite Filter Medium

The preparation of a composite filter medium was conducted on a pilot-scale customized paper machine in Fibrway Co., Ltd. (Guangzhou, China). The experimental results on the paper machine were more applicable for real applications compared to paper preparation in the laboratory [29]. During laboratory processes, water flows perpendicularly to a fixed forming screen, while for an industrial paper machine, water flows to a moving forming screen at a certain angle, which results in a different fiber orientation and structure.

The main preparation processes included pulp preparation, sheet forming, drying, and coating. There were two sets of the pulp preparation system: Microfiber and submicro-fiber slurry. For the microfiber slurry, the fibers consisted of 20% softwood pulp (Weyerhaeuser Co., Seattle, WA, USA), 60% flash-dried hardwood pulp (Suzano, Sao Paulo, Brazil), 15% 1.4 D × 5 mm polyethylene terephthalate (PET, from Kuraray Co., Ltd., Tokyo, Japan) fiber, and 5% 3-μm glass fiber was dispersed in clean water and then stored in a tank with continuous agitation to keep fibers away from entanglement. The microfiber slurry with a pulp consistency of 0.1 wt.% was pumped to channel 1 of the headbox, as shown in Figure 1. For the submicro-fiber slurry, glass wool fibers (Dongxiang Glass Microfiber Co., Ltd., Shenyang, China) with an average diameter of 0.6 μm were dispersed in water with a pH of 2.5 [30] and then stored in a tank with continuous agitation. The submicro-fiber slurry with a consistency of 0.02 wt.% was pumped to channel 2 of the headbox, as shown in Figure 1. The mass ratio of the microfiber to submicro-fiber layer was 10:1.

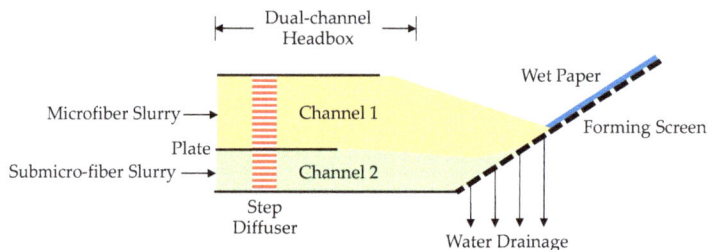

Figure 1. Schematic diagram of dual-channel headbox.

This dual-channel headbox had been patented by the authors [31]. In the dual-channel headbox, fluids in both channels were stabilized by a step diffuser and then became a steady laminar flow. These two channels were separated by a plate so that the submicro-fiber slurry did not mix with the microfiber slurry. Before entering the water drainage section, the slurries in the two channels converged at the end of the plate and were kept separated without mixing due to the steady laminar flow. Water from the submicro-fiber slurry drained first, and then the submicro-fiber deposited on the forming screen. The forming screen was a multilayer wire of PET and nylon fiber strands with a top layer pore size of 100 μm. After the submicro-fiber layer formed, microfibers started to deposit on the submicro-fiber and the dual-layer paper sheet was formed. The crucial part of the drainage section was to keep the water drainage process steady and prevent the dual-layer structure from being damaged. To achieve this goal, the submicro-fiber layer was put underneath the microfiber layer because the submicro-fiber deposition on the forming screen can slow down the initial water-draining rate and thus, make the water drainage steadier.

The wet paper sheet was then transferred to the drying and coating section, as shown in Figure 2. The dried paper from the first drying section was transferred to the gap of two resin-coating rolls. The lower roll brought 20 wt.% waterborne acrylic polymer (PR-36, Wuwei Co., Ltd., Guangzhou, China) from the resin tank to saturate the paper. This polymer can enhance the strength of the paper, especially the weak submicro-fiber layer. The second drying section was used to dry the coated paper. A major problem that existed during the drying of the coated porous material was that the resin could migrate from the inner part of paper to the surface when heated [32] and gradually clog the surface pores. Pore clogging can remarkably reduce the filtration performance and cancel out the advantages of the submicro-fiber. To mitigate binder migration, the temperature of each dryer in the second drying section was accurately controlled. The cylinder surface temperatures of dryer 4, 5, and 6 were $75 \pm 5\,°C$, $105 \pm 5\,°C$, and $125 \pm 5\,°C$, respectively. The resin content in the final dried paper was 20%. After drying and winding, the preparation of the submicro-fiber medium was complete.

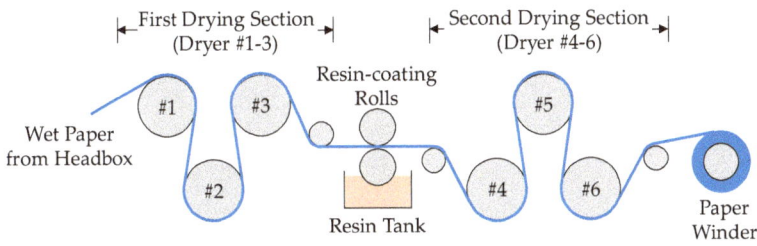

Figure 2. Schematic diagram of drying and coating process.

The running speed of the pilot-scale paper machine was 5 m/min. The basic weight of the filter paper was 120 ± 3 g/m^2. The full width of the final paper sheet was 0.7 m after cutting 0.05 m on both edges. Filter paper samples for this study were taken 1 h after stable operation of the paper machine was achieved.

2.2. Loading Performance Evaluation in Laboratory—Filter Media Specifications

The loading performance of the wet-laid submicro-fiber composite medium was compared to the two other filter media. The standard heavy-duty filter medium (H&V Inc., East Walpole, MA, USA) is currently used in many inlet filters of engineering machinery. The electrospun composite medium (Fibrway Co., Ltd., Guangzhou, China) is known for its outstanding performance in many studies [19,20]. The submicro-fiber in the electrospun medium was made from nylon-66 (Elmarco, Liberec, Czech Republic). The basic properties of these filter media are described in Table 1. The micro-structure of these filter media was observed by scanning electron microscope (Phenom^{TM}, Phenom-World B.V. Inc., Eindhoven, The Netherland). The pore size was measured by capillary flow porometer (model: CFP-1100A, Porous Materials Inc., Ithaca, NY, USA).

Table 1. Physical properties of wet-laid composite medium and two other filter media.

Parameters	Test Data			Test Standard
	Wet-Laid Composite Medium	Standard Heavy-Duty Filter Medium	Electrospun Composite Medium	
Basis weight (g/m^2)	119	110	118	TAPPI T410 [33]
Thickness (mm)	0.37	0.36	0.32	TAPPI T411 [34]
Air permeability (L/(m^2·s))	137	106	243	ASTM D737 [35]
Gravimetric efficiency (@2000 Pa) (%)	99.9	99.8	99.9	ISO 5011 [36]
Pressure drop (@11.1 cm/s) (Pa)	162	209	91	ISO 5011 [36]
Stiffness (machine direction) (mN·m)	3.0	3.7	2.6	TAPPI T489 [37]

2.3. Loading Test Setup in Laboratory

The loading test rig (MFP 3000, Palas GmbH, Karlsruhe, Germany) was used to test the dust holding capacity according to the ISO 5011 [36] standard, as shown in Figure 3.

The particles used in the test were ISO 12103-1 [38] A2 fine dust (Powder Technology Inc., Arden Hills, MN, USA). The volume mean diameter of A2 fine dust was approximately 10 μm, which was suitable to evaluate the filter medium for the engine. When the loading test started, dry A2 dust was dispersed by compressed air (0.6 MPa), and then neutralized to achieve Boltzman charge equilibrium by an ion stream from a high-voltage neutralizer. Make-up air was mixed with particles in the mixing chamber to achieve a final particle mass concentration of 1.0 mg/m^3. The filter medium clamped in the filter holder with a test area of 100 cm^2 was challenged by the dust particles under a constant flow rate controlled by a mass flow controller. The testing face velocity was 11.1 cm/s, which was close to the operating velocity in real applications [39]. The test stopped after the pressure drop reached 2 kPa. During the test, the temperature was 21 °C and the relative humidity (RH) was 50 ± 10%. For every sample, the loading test was repeated 5 times.

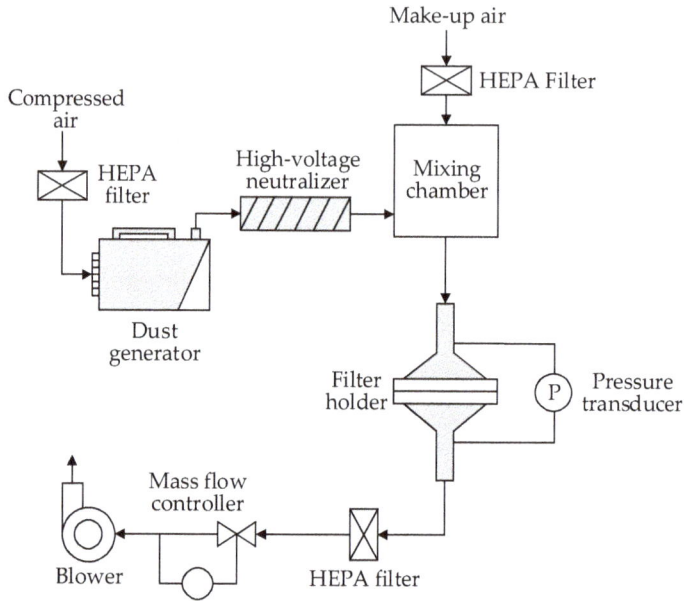

Figure 3. Schematic diagram of dust loading test setup.

The filter medium was weighed when it was clean (M_0) and fully loaded (M_1). The final dust mass deposit, M_{final}, was calculated by Equation (1).

$$M_{final} = \frac{(M_1 - M_0)}{S}$$ (1)

where S is the test area of filter medium.

Because the gravimetric efficiency for all filter media was close to 100%, as shown in Table 1, it was considered that all particles were deposited on the filter media. Dust mass deposit per unit area, M_i, at loading time of t_i can be calculated by Equation (2).

$$M_i = \frac{t_i}{t_{total}} \times M_{final}$$ (2)

where t_{total} is the total loading time (min).

The curves of pressure drop versus dust mass deposit per unit area were used to analyze the loading performance of the different media. Two parameters obtained from the pressure drop curves were considered for discussion: Dust holding capacity and energy consumption. Dust holding capacity was defined as the dust mass deposit per unit area when the pressure drop reached 2 kPa. The energy consumption of different filter media under a fixed amount of dust mass deposit can be compared by calculation methods described in Eurovent 4/21 [9].

The loading performance of the wet-laid submicro-fiber composite medium and two commercial filter media was evaluated in the laboratory. Because the wet-laid submicro-fiber composite medium was a dual-layer structure, loading tests using the microfiber or submicro-fiber layer at the inlet were conducted to analyze the difference in filtration performance, as shown in Figure 4.

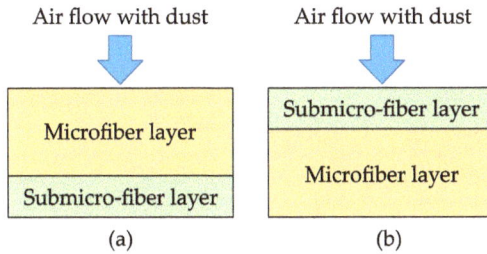

Figure 4. Loading tests of wet-laid submicro-fiber composite medium using: (**a**) microfiber layer as inlet and (**b**) submicro-fiber layer as inlet.

2.4. Loading Performance Evaluation in Field Test

The loading performance of the wet-laid submicro-fiber composite medium and the standard heavy-duty filter medium were evaluated in the field test. Because one sample of the electrospun composite medium was not big enough to make a full-size filter, no field test was conducted for the electrospun medium. During the field test, two heavy-duty trucks (Tianlong 375, Dongfeng Motor Corp., Wuhan, China) with close mileage were used. The original equipment manufacturer (OEM) air filter (Model: 3048u, Shanghai Fleetguard Filter Co., Ltd., Shanghai, China) for this truck was made by the standard heavy-duty filter medium. In comparison, a filter based on identical filter geometry was made using the wet-laid submicro-fiber medium, as shown in Figure 5.

Figure 5. OEM inlet air filter (**a**) and new submicro-fiber filter (**b**).

The two trucks equipped with the OEM filter and submicro-fiber filter operated at the same coal mine (Lyuliang Coal Mine, Shanxi, China) under the same tasks. The surrounding region of this coal mine was semi-arid, and many roads are unpaved and dusty, as shown in Figure 6, which was a harsh environment for the air filter. The nominal flow rate of air intake filter in the truck was 1500 m^3/h. After running 10,000 km, these two filters were taken out of the trucks and put into an ISO 5011 [36] test rig to conduct the pressure drop measurement under an airflow rate ranging from 750 to 1800 m^3/h. The pressure drop results were used to evaluate the loading performance of the wet-laid submicro-fiber medium and the standard heavy-duty filter medium.

Figure 6. Dusty conditions of field test site.

3. Results and Discussion

3.1. Properties and Structural Analysis of Wet-Laid Submicro-Fiber Medium

The properties of the wet-laid submicro-fiber medium are in Table 1. The wet-laid submicro-fiber medium had a close basic weight and gravimetric efficiency compared to the other two media. The initial pressure drop of the wet-laid submicro-fiber medium was 29% lower than that of the standard heavy-duty medium, but 44% higher than that of the electrospun medium. Although the initial pressure drop can indicate initial performance, it did not represent the overall performance of the loading, which is discussed in the following sections. The bending stiffness in the machine direction, which indicated the ability of sustaining the pleat structure in the operation, was very close for the three different filter media.

Figure 7 shows scanning electron microscopy (SEM) images of the wet-laid submicro-fiber medium. The microfiber and submicro-fiber layer were bonded together to create a dual-layer structure. Due to the intrinsic stiffness of glass, the glass wool fiber in the submicro-fiber layer was relatively straight and stiff, thus sustaining a highly porous fiber network. Few pores were clogged by resin in either the submicro-fiber or microfiber layer, which indicated that the temperature profile of the second drying section was an effective strategy to mitigate binder migration.

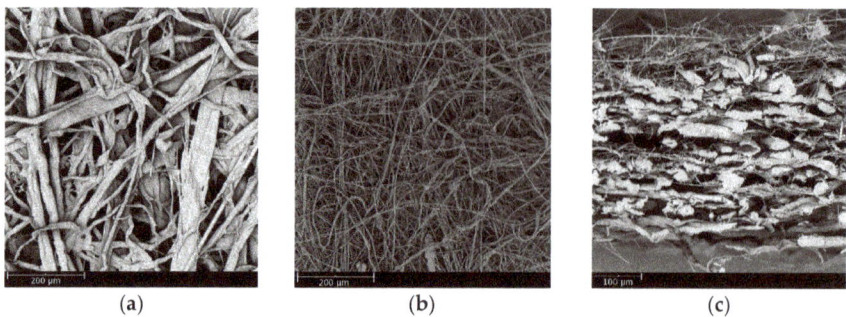

(a) (b) (c)

Figure 7. SEM images of wet-laid submicro-fiber medium: (**a**) microfiber layer, (**b**) submicro-fiber layer, and (**c**) cross-section.

3.2. Loading Performance of Wet-laid Submicro-Fiber Medium in Laboratory

When a dual-layer filter medium was used for filtration, it was important to determine which media was the inlet. Several researchers have studied the loading difference of the microfiber and submicro-fiber layers. Leung and Hung [40], Leung et al. [41], and Tang et al. [42] predicted that because the surface loading was taking place on the submicro-fiber layer, the pressure drop across the submicro-fiber filter could rise at a much faster rate than the microfiber filter. It was also hypothesized that the microfiber layer could be placed upstream in a dual-layer filter to help collect parts of the particles, thus suppressing the increased rate of pressure drop. However, their studies focused on fine particles (e.g., PM$_{2.5}$), and it must be noted that the loading behavior of coarse particles was very different. Figure 8 shows the loading results of the wet-laid submicro-fiber medium with the microfiber or submicro-fiber layer as the inlet.

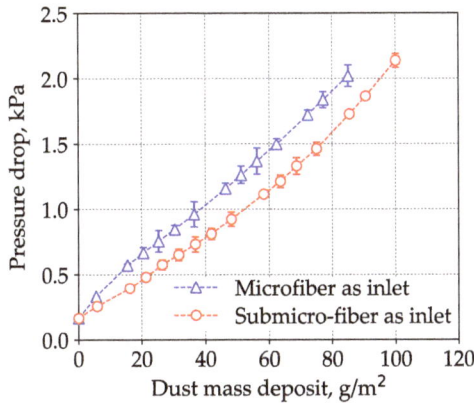

Figure 8. Loading results of wet-laid submicro-fiber medium with the microfiber or submicro-fiber layer as inlet.

The loading results demonstrated that when the submicro-fiber layer was placed as the inlet, the pressure drop increased at a lower rate. When the microfiber layer was the inlet, the dust holding capacity was 84 ± 2 g/m^2. When the submicro-fiber layer was the inlet, the dust holding capacity was 95 ± 1 g/m^2, which was 13% higher than that of the microfiber layer as the inlet. This meant that when the filter medium was challenged by dust particles, the lifetime was 13% longer if the submicro-fiber layer was placed in the upstream. This conclusion was contradictory to the conclusion of Leung et al. [41], which indicated that the loading behavior of coarse particles was different from that of fine particles.

The pressure drop of the loaded filter medium can be considered as the sum of the pressure drop across the loaded filter medium and the pressure drop of the dust cake, as shown in Figure 9.

According to Endo et al. [43,44], the pressure drop of dust cake can be calculated by Equation (3). It can be seen from Equation (3) that the pressure drop of dust cake was related to the properties of fluid and dust, and independent of filter medium. The latter part of the two curves in Figure 8 showed a similar increasing rate, which demonstrated that the pressure drop of the dust cake had no dependency on the filter medium. Both theoretical model and experimental results indicated that it was the particles' deposition pattern inside the filter medium that determined the pressure drop evolution in the loading. Therefore, the interaction between the particles and the fibrous structure in the early loading stage was the key to analyzing the results in Figure 8.

$$\Delta P_{cake} = 180 \mu u_s H \frac{(1-\varepsilon)^2}{\varepsilon^3} \frac{\kappa}{d_{vg}^2 \exp(4\ln^2 \sigma_g)} \qquad (3)$$

where ΔP_{cake} is pressure drop across the dust cake; μ is gas viscosity; u_s is face velocity; H is cake height; ε is cake porosity; κ is dynamic shape factor of A2 dust; d_{vg} is geometric mean of volume equivalent diameter; σ_g is geometric standard deviation.

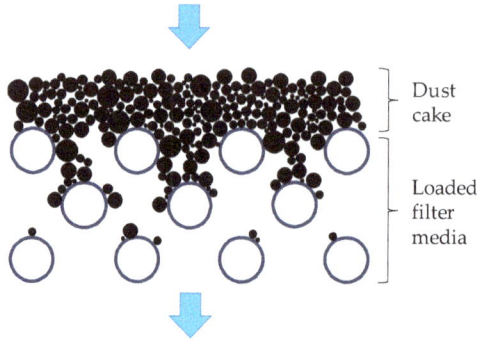

Figure 9. Schematic of dust cake and loaded filter medium.

To illustrate the loading behavior of the particles, SEM images of dust deposition on the filter medium were obtained when the pressure drop increased 10, 50, and 100 Pa, respectively, as shown in Figures 10 and 11. When the microfiber layer was placed as the inlet, dust particles penetrated the depth of the filter medium. The deposited particles bridged the pores and gradually clogged the pores. The dust particles captured inside the filter medium reduced the porosity and led to a higher pressure drop. When the submicro-fiber layer was placed as the inlet, less dust particles penetrated the depth of the filter medium and surface loading occurred rapidly. Based on the results in Figure 8, the filter medium clogged by dust particles was more difficult for air to permeate than the dust cake. Therefore, it was preferable for surface loading to happen earlier for coarse particles, while in the studies of Leung et al. [41] and Tang et al. [42], surface loading should take place later when fine particles were tested. For this reason, the submicro-fiber layer was placed as the inlet in the following discussions.

(a) (b) (c)

Figure 10. SEM images of dust deposition when microfiber layer was placed as inlet and pressure drop increase by: (**a**) 10 Pa, (**b**) 50 Pa, and (**c**) 100 Pa.

Figure 11. SEM images of dust deposition when submicro-fiber layer was placed as inlet and pressure drop increase by: (**a**) 10 Pa, (**b**) 50 Pa, and (**c**) 100 Pa.

3.3. Loading Performance Comparison Between Wet-laid Submicro-Fiber Medium and Standard Heavy-duty Medium in the Laboratory and Field

The loading results of the wet-laid submicro-fiber medium and the standard heavy-duty medium are shown in Figure 12. The pressure drop of the standard heavy-duty medium increased faster than that of the wet-laid submicro-fiber medium. When the pressure drop reached 2 kPa, the dust mass deposit of the standard heavy-duty medium was 64 g/m^2, which meant that the dust holding capacity of the wet-laid submicro-fiber medium was 48% higher than that of the standard heavy-duty medium.

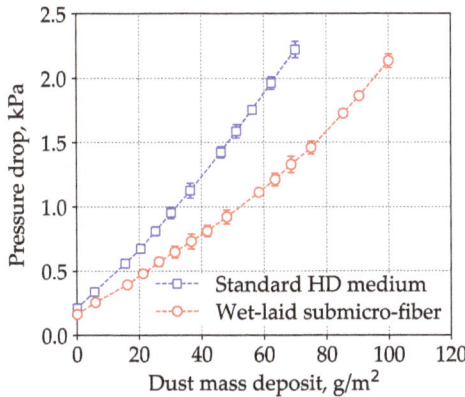

Figure 12. Pressure drop curves of standard heavy-duty (HD) medium and wet-laid submicro-fiber medium in the loading process.

The energy consumption of air filter media was proportional to average pressure drop in loading process [9]. $\Delta p(M_i)$ was the pressure drop when dust mass deposit was M_i. The average pressure drop in a complete loading test can be calculated by Equation (4). The polynomial of fourth order $\Delta p(m) = a_n M_i^4 + b_n M_i^3 + c_n M_i^2 + d_n M_i + \Delta p_I$ and parameters a_n, b_n, c_n and d_n were used to curve fit the pressure drop data as a function of loading time. Δp_I is the initial pressure drop of filter media.

$$\overline{\Delta p} = \frac{1}{M_{final}} \int_0^{M_{final}} \Delta p(M_i) dM_i \tag{4}$$

Under a dust mass deposit of 64 ± 1 g/m^2, the estimated energy consumption of the wet-laid submicro-fiber medium was 36% lower than that of standard heavy-duty medium based on calculation of average pressure drop via Equation (4). The laboratory results showed that the wet-laid submicro-fiber medium had a longer lifetime and lower energy consumption than the standard heavy-duty medium when challenged by dust particles.

The SEM images of the clean standard heavy-duty medium and the loaded medium when the pressure drop increased 10, 50, and 100 Pa were obtained, respectively, as shown in Figure 13. The dust particles penetrated into the depth of the fibrous structure. Surface loading took place later than the wet-laid submicro-fiber medium compared to Figure 11. The dust loading behavior was similar to that of the microfiber layer in Figure 10. The results further confirmed that depth filtration was not suitable for coarse particle filtration.

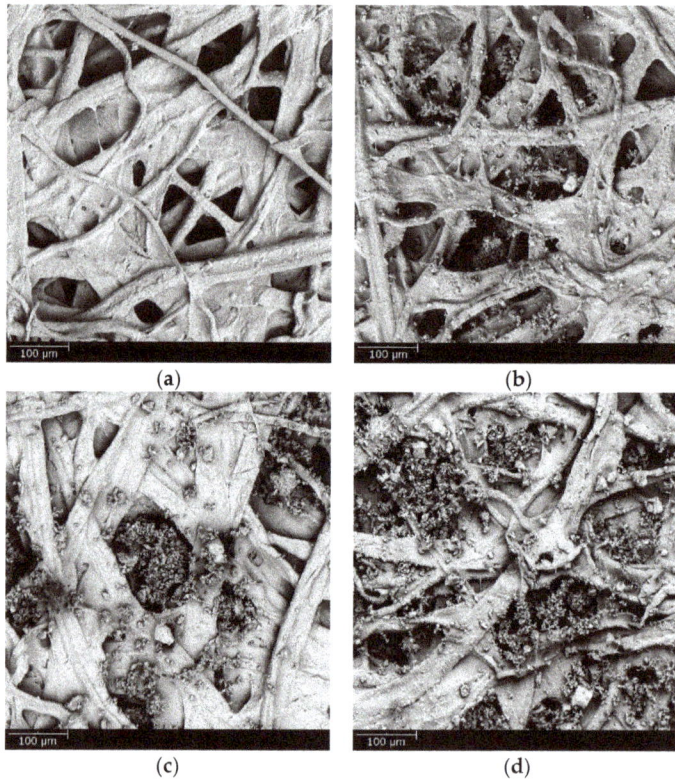

Figure 13. SEM images of dust deposition of standard heavy-duty filter medium: (**a**) clean medium, (**b**) pressure drop increase of 10 Pa, (**c**) pressure drop increase of 50 Pa, and (**d**) pressure drop increase of 100 Pa.

Figure 14 shows the pressure drop of the standard heavy-duty filter and the wet-laid submicro-fiber filter under different flow rates after 10,000 km of operation. Under the nominal flow rate of the air filter, 1500 m^3/h, the pressure drop of the standard heavy-duty filter was higher than 2 kPa and the filter service life ended. The pressure drop of the used wet-laid submicro-fiber filter was approximately 45% lower than that of the standard heavy-duty filter, which demonstrated the advantage of the novel filter medium in the field test. During the laboratory tests, when the pressure drop of the standard heavy-duty medium reached 2 kPa, the pressure drop of the wet-laid submicro-fiber medium under the same dust mass deposit was 38% lower than that of the standard heavy-duty medium, which was close to the field test results. The field test results showed that the wet-laid submicro-fiber medium had a longer lifetime when challenged by dust particles.

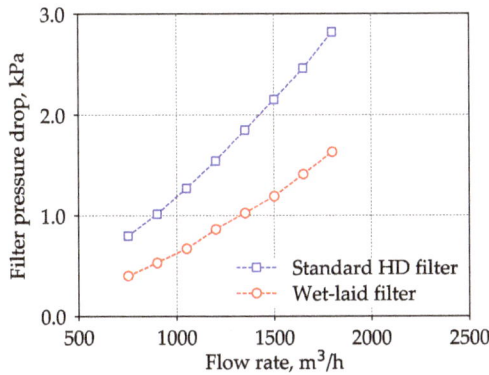

Figure 14. Pressure drop of used standard heavy-duty (HD) filter and wet-laid submicro-fiber filter under different flow rates after running 10,000 km.

3.4. Loading Performance Comparison Between Wet-Laid Submicro-Fiber Medium and Electrospun Composite Medium in Laboratory

The loading results of the wet-laid submicro-fiber medium and an electrospun composite medium are shown in Figure 15. The initial pressure drop of the electrospun medium was lower, but the pressure drop increased faster than the wet-laid submicro-fiber medium. When the pressure drop reached 2 kPa, the dust mass deposit of the electrospun medium was 86 ± 1 g/m^2, which meant that the dust holding capacity of the wet-laid submicro-fiber medium was 10% higher than that of the electrospun medium. Under a dust mass deposit of 86 g/m^2, the estimated energy consumption of the wet-laid submicro-fiber medium was 10% lower than that of the electrospun medium based on calculation of average pressure drop via Equation (4). The laboratory results showed that the wet-laid submicro-fiber medium had a longer lifetime and lower energy consumption than the electrospun medium when challenged by dust particles.

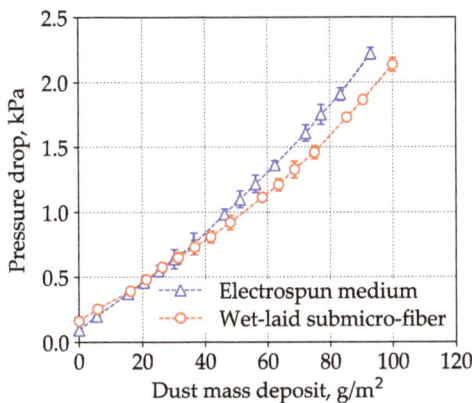

Figure 15. Pressure drop curves of electrospun composite medium and wet-laid submicro-fiber medium in the loading process with submicro-fiber as upstream.

The SEM images of the clean electrospun composite medium and loaded filter medium when the pressure drop increased 10, 50, and 100 Pa were obtained, respectively, as shown in Figure 16. The electrospun medium was believed to be a typical surface loading material. When the pressure drop increased 10 kPa, the electrospun medium had already shown evident signs of surface loading. Although surface loading was preferred in the removal of coarse particles, the process of surface loading would affect the loading performance.

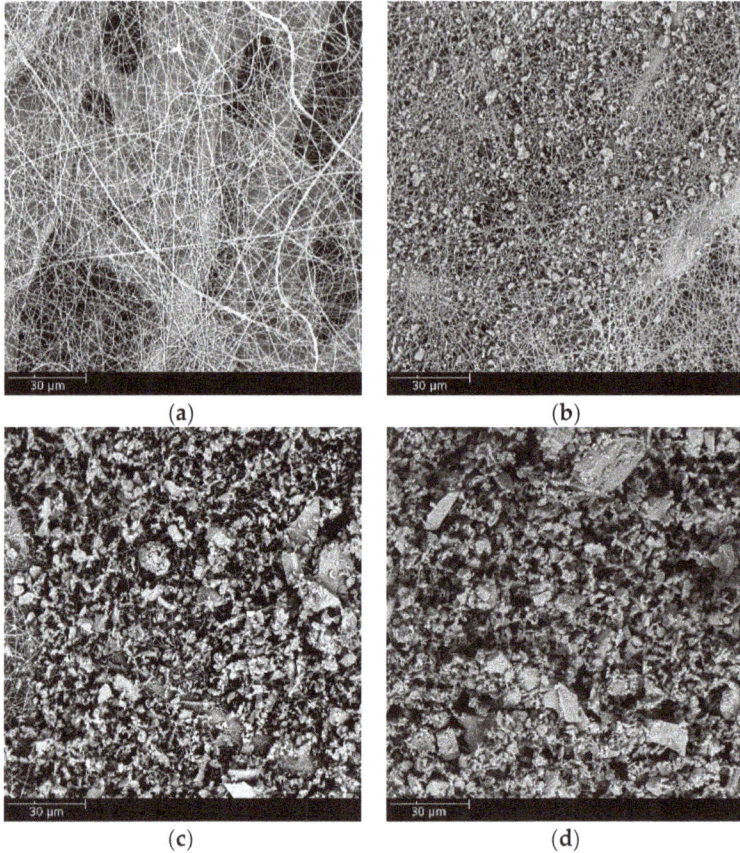

Figure 16. SEM images of dust deposition of electrospun composite medium: (**a**) clean medium, (**b**) pressure drop increase of 10 Pa, (**c**) pressure drop increase of 50 Pa, and (**d**) pressure drop increase of 100 Pa.

The governing mechanisms of capturing coarse particles were interception, inertial impaction, and sieving [14]. Particles were captured by fibers due to interception and inertial impaction, while particles were arrested at the pores (smaller than the particle diameter) due to sieving. Figure 11 shows that the dust particles were largely captured by submicro-fibers of the wet-laid medium, while particles were largely arrested by pores of the electrospun medium as shown in Figure 16. Figure 17 was the pore size distribution of electrospun media and wet-laid submicro-fiber media. The peak pore size of electrospun media and wet-laid submicro-fiber media was 3.3 μm and 8.8 μm, respectively. The pore size distribution of electrospun media was narrower than that of wet-laid submicro-fiber media. Since the volume mean diameter of A2 fine dust was approximately 10 μm [38], more dust particles were captured by pores instead of fibers for electrospun media than wet-laid submicro-fiber media. When the pores were blocked by larger particles, the airflow path through the fibers and particles was restricted. Airflow velocity in the restricted pores rapidly increased and thus the pressure drop increased. The unique fibrous structure of the wet-laid submicro-fiber medium made more particle deposition take place on fibers via interception and inertial impaction rather than sieving, which led to a porous structure with a larger airflow path. Therefore, the wet-laid submicro-fiber medium showed a better loading performance against coarse particles than the electrospun medium.

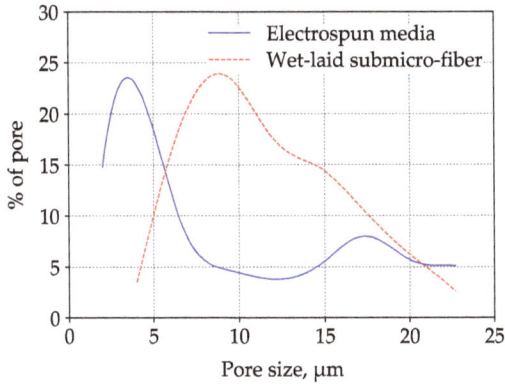

Figure 17. Pore size distribution of electrospun media and wet-laid submicro-fiber media.

The current study is limited to room temperature and controlled RH (50 ± 10%). In the future, the effect of excess RH and temperature on dust loading performance of wet-laid submicro-fiber media should be studied.

4. Conclusions

In this study, a novel wet-laid submicro-fiber composite medium was prepared by a dual-layer paper machine, and its loading performance and dust deposition behavior was compared with a standard heavy-duty medium and electrospun medium in laboratory and field tests. It was found that:

(1) The submicro-fiber layer, laminated by dual-channel headbox technology and reinforced by acrylic resin, was highly porous without pore clogging. The loading results demonstrated that when the submicro-fiber layer was placed as the inlet, the pressure drop increased at a slower rate than that when the microfiber layer was the inlet. It was concluded that it was preferable for surface loading to happen earlier for dust particles.

(2) During the laboratory test, the dust holding capacity of the wet-laid submicro-fiber medium was 48% higher than that of the standard heavy-duty medium, and the estimated energy consumption was 36% lower. In the field test, the pressure drop of the wet-laid submicro-fiber filter was approximately 45% lower than that of the standard heavy-duty filter after 10,000 km of operation. This showed that the wet-laid submicro-fiber medium had a longer lifetime and lower energy consumption than the standard heavy-duty medium when challenged by dust particles.

(3) The dust holding capacity of the wet-laid submicro-fiber medium was 10% higher than that of the electrospun medium, and the estimated energy consumption was 10% lower. The key point of submicro-fiber medium design was that the particles were preferable to be captured by fibers rather than pores. The fibrous structure of the wet-laid submicro-fiber medium made more particle deposition take place on fibers via interception and inertial impaction. Thus, in this work, the submicro-fiber medium showed a lower pressure drop than the electrospun medium in the loading process.

Author Contributions: Conceptualization, J.L. and M.T.; methodology, J.L.; software, M.T.; validation, J.L.; formal analysis, J.L. and M.T.; investigation, J.L. and Z.S.; resources, Y.L.; data curation, J.L.; writing (original draft preparation), J.L.; writing (review and editing), M.T. and Y.L.; supervision, J.H.; project administration, Y.L.; funding acquisition, J.H.

Funding: This research was funded by the National Key R and D Program of China, Grant No. 2017YFB0308000.

Conflicts of Interest: The authors declare no conflict of interest.

References

1. Chun, Y.; Kim, J.; Choi, J.C.; Boo, K.O.; Oh, S.N.; Lee, M. Characteristic number size distribution of aerosol during Asian dust period in Korea. *Atmos. Environ.* **2001**, *35*, 2715–2721. [CrossRef]
2. Csavina, J.; Field, J.; Taylor, M.P.; Gao, S.; Landázuri, A.; Betterton, E.A.; Eduardo, S.A.A. Review on the importance of metals and metalloids in atmospheric dust and aerosol from mining operations. *Sci. Total Environ.* **2012**, *433*, 58–73. [CrossRef] [PubMed]
3. Lakshminarayanan, P.A.; Nayak, N.S. Wear in the heavy duty engine. In *Critical Component Wear in Heavy Duty Engines*, 1st ed.; John Wiley & Sons (Asia) Pte Ltd.: Singapore, 2011; pp. 1–11. ISBN 9780470828823.
4. Sherburn, P.E. Air cleaner design—Present and future. *SAE Tech. Pap.* **1969**, 690007. [CrossRef]
5. Treuhaft, M.B. The use of radioactive tracer technology to measure engine ring wear in response to dust ingestion. *SAE Tech. Pap.* **1993**, 930019. [CrossRef]
6. Jaroszczyk, T.; Petrik, S.; Donahue, K. Recent development in heavy duty engine air filtration and the role of nanofiber filter media. *J. Kones* **2009**, *16*, 207–216.
7. Schilling, A. *Automobile Engine Lubrication*; Scientific Publications (G.B.) Limited: Broseley, UK, 1972; ISBN 9780900645044.
8. Wilcox, M.; Baldwin, R.; Garcia-Hernandez, A.; Brun, K. *Guideline for Gas Turbine Inlet Air Filtration Systems*; Gas Machinery Research Council: Dallas, TX, USA, 2010.
9. Calculation Method for the Energy Use Related to Air Filters in General Ventilation Systems, Eurovent 4/21. Available online: https://eurovent.eu/?q=content/eurovent-421-2014-calculation-method-energy-use-related-air-filters-general-ventilation (accessed on 15 October 2018).
10. Long, J.; Tang, M.; Liang, Y.; Hu, J. Preparation of fibrillated cellulose nanofiber from lyocell fiber and its application in air filtration. *Materials* **2018**, *11*, 1313. [CrossRef] [PubMed]
11. Podgórski, A.; Bałazy, A.; Gradoń, L. Application of nanofibers to improve the filtration efficiency of the most penetrating aerosol particles in fibrous filters. *Chem. Eng. Sci.* **2006**, *61*, 6804–6815. [CrossRef]
12. Zhang, Q.; Welch, J.; Park, H.; Wu, C.Y.; Sigmund, W.; Marijnissen, J.C.M. Improvement in nanofiber filtration by multiple thin layers of nanofiber mats. *J. Aerosol Sci.* **2010**, *41*, 230–236. [CrossRef]
13. Hung, C.H.; Leung, W.F. Filtration of nano-aerosol using nanofiber filter under low peclet number and transitional flow regime. *Sep. Purif. Technol.* **2011**, *79*, 34–42. [CrossRef]
14. Wang, C.; Otani, Y. Removal of nanoparticles from gas streams by fibrous filters: A review. *Ind. Eng. Chem. Res.* **2013**, *52*, 5–17. [CrossRef]
15. He, Z.; Zhang, X.; Batchelor, W. Cellulose nanofibre aerogel filter with tuneable pore structure for oil/water separation and recovery. *RSC Adv.* **2016**, *6*, 21435–21438. [CrossRef]
16. Si, Y.; Fu, Q.; Wang, X.; Zhu, J.; Yu, J.; Sun, G.; Ding, B. Superelastic and superhydrophobic nanofiber-assembled cellular aerogels for effective separation of oil/water emulsions. *ACS Nano* **2015**, *9*, 3791–3799. [CrossRef] [PubMed]
17. Esfahani, H.; Jose, R.; Ramakrishna, S. Electrospun Ceramic Nanofiber Mats Today: Synthesis, Properties, and Applications. *Materials* **2017**, *10*, 1238. [CrossRef] [PubMed]
18. Yuan, W.; Fang, G.; Li, Z.; Chen, Y.; Tang, Y. Using electrospinning-based carbon nanofiber webs for methanol crossover control in passive direct methanol fuel cells. *Materials* **2018**, *11*, 71. [CrossRef] [PubMed]
19. Choi, J.; Yang, B.J.; Bae, G.N.; Jung, J.H. Herbal extract incorporated nanofiber fabricated by an electrospinning technique and its application to antimicrobial air filtration. *ACS Appl. Mater. Inter.* **2015**, *7*, 25313–25320. [CrossRef] [PubMed]
20. Chang, C.Y.; Chang, F.C. Development of electrospun lignin-based fibrous materials for filtration applications. *BioResources* **2016**, *11*, 2202–2213. [CrossRef]
21. Barhate, R.S.; Ramakrishna, S. Nanofibrous filtering media: Filtration problems and solutions from tiny materials. *J. Membr. Sci.* **2007**, *296*, 1–8. [CrossRef]

22. Saleem, M.; Gernot, K. Effect of filtration velocity and dust concentration on cake formation and filter operation in a pilot scale jet pulsed bag filter. *J. Hazard. Mater.* **2007**, *144*, 677–681. [CrossRef] [PubMed]
23. Zhang, S.; Liu, H.; Yin, X.; Yu, J.; Ding, B. Anti-deformed polyacrylonitrile/polysulfone composite membrane with binary structures for effective air filtration. *ACS Appl. Mater. Interfaces* **2016**, *8*, 8086–8095. [CrossRef] [PubMed]
24. Wang, N.; Yang, Y.; Al-Deyab, S.S.; El-Newehy, M.; Yu, J.; Ding, B. Ultra-light 3D nanofibre-nets binary structured nylon 6–polyacrylonitrile membranes for efficient filtration of fine particulate matter. *J. Mater. Chem.* **2015**, *A3*, 23946–23954. [CrossRef]
25. Wang, N.; Si, Y.; Wang, N.; Sun, G.; El-Newehy, M.; Al-Deyab, S.S.; Ding, B. Multilevel structured polyacrylonitrile/silica nanofibrous membranes for high-performance air filtration. *Sep. Purif. Technol.* **2014**, *126*, 44–51. [CrossRef]
26. Leung, W.W.F.; Hau, C.W.Y.; Choy, H.F. Microfiber-nanofiber composite filter for high-efficiency and low pressure drop under nano-aerosol loading. *Sep. Purif. Technol.* **2018**, *206*, 26–38. [CrossRef]
27. Langner, M.; Greiner, A. Wet-laid meets electrospinning: Nonwovens for filtration applications from short electrospun polymer nanofiber dispersions. *Macromol. Rapid. Commun.* **2016**, *37*, 351–355. [CrossRef] [PubMed]
28. Galka, N.; Saxena, A. High efficiency air filtration: The growing impact of membranes. *Filtr. Sep.* **2009**, *46*, 22–25. [CrossRef]
29. Yao, Y.; Tang, M.; Yu, T.; Liang, Y.; Hu, J. Filtration performance of dual-layer filter paper with fibrillated nanofibers. *BioResources* **2016**, *11*, 9506–9519. [CrossRef]
30. Tang, M.; Hu, J.; Liang, Y.; Pui, D.Y.H. Pressure drop, penetration and quality factor of filter paper containing nanofibers. *Text. Res. J.* **2017**, *87*, 498–508. [CrossRef]
31. Hu, J.; Liang, Y.; Wang, Y.; Zeng, J. Self-cleaning air filtering material and preparation method therefor. U.S. Patent 9,771,904, 26 September 2017.
32. Bernada, P. Experimental study and modelling of binder migration during drying of a paper coating. *Dry. Technol.* **1996**, *14*, 1897–1899. [CrossRef]
33. *Grammage of Paper and Paperboard (Weight per Unit Area)*; TAPPI T410 om-13; TAPPI Press: Atlanta, GA, USA, 2013.
34. *Thickness (Caliper of Paper, Paperboard, and Combined Board)*; TAPPI T411 om-15; TAPPI Press: Atlanta, GA, USA, 2015.
35. *Standard Test Method for Air Permeability of Textile Fabrics*; ASTM D737-18; ASTM International: West Conshohocken, PA, USA, 2018.
36. ISO. *Inlet Air Cleaning Equipment for Internal Combustion Engines and Compressors—Performance Testing*; ISO 5011; International Organization for Standardization: Geneva, Switzerland, 2014; Available online: https://www.iso.org/obp/ui/#iso:std:iso:5011:ed-3:v1:en (accessed on 15 October 2018).
37. *Bending Resistance (Stiffness) of Paper and Paperboard (Taber-Type Tester in Basic Configuration)*; TAPPI T489 om-15; TAPPI Press: Atlanta, GA, USA, 2015.
38. ISO. *Road Vehicles—Test Contaminants for Filter Evaluation—Part 1: Arizona Test Dust*; ISO 12103-1; International Organization for Standardization: Geneva, Switzerland, 2016; Available online: https://www.iso.org/obp/ui/#iso:std:iso:12103:-1:ed-2:v1:en (accessed on 15 October 2018).
39. Sun, Z.; Tang, M.; Song, Q.; Yu, J.; Liang, Y.; Hu, J.; Wang, J. Filtration performance of air filter paper containing kapok fibers against oil aerosols. *Cellulose* **2018**, *4*, 1–11. [CrossRef]
40. Leung, W.W.F.; Hung, C.H. Investigation on pressure drop evolution of fibrous filter operating in aerodynamic slip regime under continuous loading of submicron aerosols. *Sep. Purif. Technol.* **2008**, *63*, 691–700. [CrossRef]
41. Leung, W.W.F.; Hung, C.H.; Yuen, P.T. Experimental investigation on continuous filtration of sub-micron aerosol by filter composed of dual-layers including a nanofiber layer. *Aerosol Sci. Technol.* **2009**, *43*, 1174–1183. [CrossRef]
42. Tang, M.; Chen, S.C.; Chang, D.Q.; Xie, X.; Sun, J.; Pui, D.Y.H. Filtration efficiency and loading characteristics of $PM_{2.5}$ through composite filter media consisting of commercial HVAC electret media and nanofiber layer. *Sep. Purif. Technol.* **2018**, *198*, 137–145. [CrossRef]

43. Endo, Y.; Chen, D.R.; Pui, D.Y.H. Effects of particle polydispersity and shape factor during dust cake loading on air filters. *Powder Technol.* **1998**, *98*, 241–249. [CrossRef]
44. Endo, Y.; Chen, D.R.; Pui, D.Y.H. Air and water permeation resistance across dust cakes on filters—Effects of particle polydispersity and shape factor. *Powder Technol.* **2001**, *118*, 24–31. [CrossRef]

materials

MDPI

Article

Preparation of Fibrillated Cellulose Nanofiber from Lyocell Fiber and Its Application in Air Filtration

Jin Long [1], Min Tang [2,*], Yun Liang [1] and Jian Hu [1]

[1] State Key Laboratory of Pulp and Paper Engineering, South China University of Technology, Guangzhou 510641, China; lj_king@139.com (J.L.); liangyun@scut.edu.cn (Y.L.); ppjhu@scut.edu.cn (J.H.)

[2] Department of Mechanical Engineering, University of Minnesota—Twin Cities, Minneapolis, MN 55455, USA

* Correspondence: tangm@umn.edu; Tel.: +1-612-840-8230

Received: 9 July 2018; Accepted: 27 July 2018; Published: 29 July 2018

Abstract: Ambient particulate matter less than 2.5 µm ($PM_{2.5}$) can substantially degrade the performance of cars by clogging the air intake filters. The application of nanofibers in air filter paper can achieve dramatic improvement of filtration efficiency with low resistance to air flow. Cellulose nanofibers have gained increasing attention because of their biodegradability and renewability. In this work, the cellulose nanofiber was prepared by Lyocell fiber nanofibrillation via a PFI-type refiner, and the influence of applying a cellulose nanofiber on filter paper was investigated. It was found that the cellulose nanofibers obtained under 1.00 N/mm and 40,000 revolutions were mainly macrofibrils of Lyocell fiber with average fiber diameter of 0.8 µm. For the filter papers with a different nanofiber fraction, both the pressure drop and fractional efficiency increased with the higher fraction of nanofibers. The results of the figure of merit demonstrated that for particles larger than 0.05 µm, the figure of merit increased substantially with a 5% nanofiber, but decreased when the nanofiber fraction reached 10% and higher. It was concluded that the optimal fraction of the cellulose nanofiber against $PM_{2.5}$ was 5%. The results of the figure of merit were related to the inhomogeneous distribution of nanofibers in the fibrous structure. The discrepancy of the theoretical and measured pressure drop showed that a higher nanofiber fraction led to a higher degree of fiber inhomogeneity.

Keywords: cellulose nanofiber; Lyocell fiber; $PM_{2.5}$; filter paper

---

1. Introduction

Clean air is considered a basic requirement for modern society. However, the air quality deterioration and the increasing air pollution has aroused widespread public concerns. The U.S. Environmental Protection Agency developed the first standard for particulate matter less than 2.5 µm ($PM_{2.5}$) in 1997, as a result of concern over the health effects of fine particles in the air [1]. Some major cities in China and India, for example, Beijing and New Delhi, are frequently exposed to severe $PM_{2.5}$ pollution. Besides the health effects, ambient particles can affect human life in other ways. One important effect is that it can substantially degrade the performance of cars, as the combustion engine of a car needs to ingest large quantities of air. The main impact of $PM_{2.5}$ on filtration performance, particularly in combination with moisture [2,3], is that it will clog the filters in a shorter time than conventional dust [4]. Fine particles are mainly collected by interception and diffusion [5], and they form dendritic particle structures that contribute to airflow resistance substantially [6]. As the initial efficiency of the traditional filter paper against fine particles is low, the $PM_{2.5}$ pollution challenges the filter performance and consequently increases the potential damage for the vehicle engines. As a result of the increasing problems of fine particle pollution, the requirement of better filter performance is more demanding. In addition, the global commitment to cleaner energy and less energy usage is motivating the filtration industry to develop a filter media with less resistance to air flow, thereby reducing energy usage.

Most of the filter media for engine intake filtration are in the form of wet-laid paper [7]. Filter paper is pleated so that the filtration area and dirt-holding capacity could be increased many times in a small space left for air filter [8]. Cellulose-type filter paper, made of fibers with diameters larger than 10 microns, are often used in engine air filtration. To improve filtration performance, the application of nanofiber has gained lots of attention. Many theoretical results and experimental data have shown that the filtration efficiency could be dramatically improved with a low resistance to air flow, due to the slip effect at the gas-fiber interface and the large specific surface area of the nanofiber [9–12]. There are two popular methods of applying nanofibers. One method is to laminate membrane-like nanofiber layer on top of the micrometer fiber by electrospinning, which will result in rapid clogging by the deposited particles, as described by Leung et al. [13]. Another method of producing the nanofiber composite is to add nanofibers (mostly glass wool fiber) in the wet-laid forming process. The nanofibers are well dispersed among the micrometer fibers, which can solve the problems of laminated nanofiber filter paper, such as weak strength and short service life. Although glass wool fiber shows an excellent filtration performance, it causes problems for the used filter paper when being disposed, as glass fiber is non-biodegradable and cannot be burned. Moreover, there is a health concern over glass wool fiber. Glass wool fibers, such as E-glass and 475 glass fibers, are considered by International Agency for Research on Cancer, IARC, as possible carcinogens to humans (Group 2B). There is a demand to find an alternative nanofiber that is human friendly and suitable for a wet-laid process.

As the search for efficient bio-based materials is the main challenge of the next decades [14], the development of cellulose nanofibers has gained increasing attention [15–23], because of their high strength, biodegradability, and renewability [24]. The extensive separation of cellulose fibers into nanofibers can be achieved if conventional refining methods are applied well beyond the levels used in paper making [25–27]. Among either natural or man-made cellulose fibers, the Lyocell fibers differ from others in its high crystallinity, high longitudinal orientation of crystallites, and low lateral cohesion between fibrils [28–30]. Lyocell is an environmentally benign man-made fiber, because it is manufactured by cellulose dissolution in nontoxic N-methyl morpholine N-oxide instead of toxic carbon disulfide, which can be almost totally recycled [31]. In the swollen state, the Lyocell fiber has an extensive tendency to fibrillate [32]. Several studies have investigated the fibrillation of the Lyocell fibers by different fibrillating equipment (e.g. metal balls with tumbling [33] and crockmeter [34]), but the properties of the fibrillated fibers that were obtained from these studies were highly dependent on their fibrillating equipment. The study of Lyocell fibrillation on a more reproducible machine (e.g. PFI-type pulp refiner [35]) was necessary. Although the fibrillated Lyocell nanofiber is a very interesting material for eco-friendly filter paper, not much literature has been reported about its application in engine filtration. Unlike other applications, the nanofiber in filter paper has its unique way of designing a fibrous structure. Tang et al. [36] and Choi et al. [37] found that adding more nanofibers did not lead to a better filtration performance, as expected by the conventional filtration theory. The study of the nanofiber structure is critical to optimize the performance of filter paper. More studies of using nanofibers in filtration are needed. The goal of this work is to prepare Lyocell cellulose nanofiber suitable for wet-laid filter paper, and to explain the microstructure changes in the fibrillation process using the PFI-type refiner. The optimization of the nanofiber mixing fraction will also be studied based on the results of the filtration test against the $PM_{2.5}$ and analysis of the fibrous structure.

2. Materials and Methods

2.1. Nanofibrillation of Lyocell Fiber

Lyocell fiber (TENCEL™, Lenzing AG, Lenzing, Austria) with a linear mass density of 1.7 dtex and length of 4 mm, as shown in Table S1, was used to prepare the cellulose nanofiber. As the Lyocell fibers were continuously drawn from the spinneret pierced with small holes in the manufacturing process, the Lyocell fibers possessed a uniform diameter (12 µm in this study) and circular cross section.

The nanofibrillation of the Lyocell fiber was conducted on a PFI-type pulp refiner (model: mark VI, Hamjern Maskjn, Hamar, Norway). The PFI refiner is a machine with a high reproducibility for use in the laboratory under standardized conditions [35]. The key elements of the PFI refiner were a stainless steel roll with chiseled bars and a cylindrical container with a smooth interior wall. The roll and the container were independently driven in the same direction, with a constant difference of peripheral speed, which resulted in mechanical effects such as shear and compression for fiber fibrillation. In this study, the refining followed the procedures in the TAPPI T248 [38] standard. The Lyocell fibers were weighed and then mixed with distilled water to form a slurry with a fiber consistency (concentration) of 10%. The fiber slurry was pressed evenly to the interior wall of the container. In the refining process, the roll and the cylindrical container rotated in the same direction at 1440 rpm (revolutions per minute) and 720 rpm, respectively. To study the impact of the refining pressure on the fibrillation, the roll was pressed against the container wall under two pressures, 1.00 N/mm and 3.33 N/mm (recommended pressure in TAPPI T248). For each refining pressure, the refining procedure was stopped and the fibrillated fiber was obtained when the revolution number of refiners reached to 5000, 10,000, 20,000, and 40,000, respectively.

2.2. Characterization of Lyocell Nanofiber

The morphology of the fibrillated Lyocell fiber was observed by an optical microscope (model: BX51TF, Olympus Corporation, Tokyo, Japan). The images were captured by a photomicrographic system equipped with the microscope, which can give a broad view to analyze the fibrillation and length of the Lyocell fiber. Based on the analysis of the microscopy images, a preferred refining pressure was selected for the subsequent studies. After determining the preferred refining pressure, the main influencing factor of the fibrillation process was the number of revolutions. The fibrillation and nanofiber morphology under different revolution numbers were observed by a scanning electron microscope (model: EV018, Carl Zeiss AG, Jena, Germany).

As the fibrillated Lyocell fiber consisted of microfiber and nanofiber, it was hard to obtain an average fiber diameter for the fibrillated Lyocell fiber. The beating degree and specific surface area were used to characterize the fiber diameter after fibrillation. The beating degree was measured by Schopper-Riegler beating tester (model: ZJG-100, Changchun Yueming Small Tester Co., Ltd, Jilin, China) following ISO 5267-1 [39]. The specific surface area was tested using the Brunauer–Emmett–Teller surface area analyzer (model: SA 3100, Beckman Coulter, Brea, CA, USA). Based on the specific surface area, the average diameter of the nanofiber, d_f (µm), can be estimated by Equation (1), as follows:

$$d_f = \frac{4000}{\rho S} \tag{1}$$

where, ρ is the fiber density, kg/m^3, and S is the specific surface area, m^2/g.

The weighted average fiber length (length-weighted) and fine fiber content were measured using the fiber quality analyzer (model: FS-300, Metso Automation, Vantaa, Finland). After the characterization and evaluation of the fibrillated Lyocell fiber, an optimal refining condition was selected and the nanofiber obtained from this condition was used to prepare the nanofiber composite filter paper.

2.3. Preparation of Nanofiber Composite Filter Paper

The fibers used in commercial filter paper for engine air intake were wood fiber and synthetic fiber. In this study, softwood fiber (from Suzano Pulp and Paper Inc., São Paulo, Brazil) and PET (polyethylene terephthalate) fiber (dimension: 1.7 dtex × 5 mm, Kuraray Co. Ltd, Tokyo, Japan) were selected to mix with the Lyocell nanofiber. The weight fraction of each fiber in the filter paper can be seen in Table 1. The basic weight for all of the samples was 105 ± 2 g/m^2.

Table 1. Weight fraction of fibers in filter paper.

Sample No.	Softwood Fiber	PET Fiber	Lyocell Nanofiber
#1	75%	25%	0
#2	70%	25%	5%
#3	65%	25%	10%
#4	60%	25%	15%
#5	55%	25%	20%

The preparation of the nanofiber composite filter paper was done according to the procedures provided by ISO 5269-1:2005 [40]. The fibers were weighed by an electronic balance with 1.0 mg accuracy, and were mixed together with 2.0 L water. The mixture was dispersed in the standard fiber disintegrator (model: P95568, PTI GmbH, Laakirchen, Austria) under 3000 rpm for 200 seconds. The well-dispersed fiber suspension was transferred to the handsheet former (model: RK3AKWT, PTI GmbH, Laakirchen, Austria). Then, 5.0 L of water was added into the handsheet former to decrease the fiber consistency and to improve the handsheet formation. After the agitation is completed, the fiber suspension was drained through a metallic screen (120 mesh). The fibers were retained on the circular screen, with a diameter of 200 mm, and thus a wet handsheet was formed. Finally, the wet handsheet was carefully transferred to the dryer (Speed Dryer, PTI GmbH, Laakirchen, Austria) and dried under 105 °C for 20 minutes. The dryer was carefully selected to minimize the compression on the porous structure of the handsheet and to keep the handsheet flat during the drying process. After the preparation of the filter media, the thickness was measured according to Tappi T411 [41], and the mean flow pore size was measured by a capillary flow porometer (model: CFP 1100, Porous Material Inc., New York, NY, USA) using a wet up/dry down procedure based on ASTM F316-03 [42]. The porosity (ε) of the filter paper can be calculated from the basic weight (BW, g/m^2), thickness (l, mm), and fiber density (ρ_f, kg/m^3), as shown in Equation (2). The preparation and testing for each type of paper were repeated five times.

$$\varepsilon = 1 - \frac{BW}{\rho_f l} \tag{2}$$

2.4. Filtration Test of Nanofiber Composite Filter Media

To evaluate the capability of capturing the PM$_{2.5}$, the fractional efficiency against the particles ranging from 30 nm to 2 μm was measured. This range covered most of the ambient PM$_{2.5}$, as indicated by the particle size distributions of Thielke et al. [43]. The fractional efficiency was measured using monodisperse potassium chloride (KCl) particles with diameter of 30 nm, 50 nm, 80 nm, 100 nm, 150 nm, and 200 nm, as described by Tang et al. [44], and polydisperse KCl particles ranging from 400 nm to 2 μm, as described in the following section.

The polydisperse KCl particles were generated by an aerosol generator (model: RBG 2000, Palas GmbH, Karlsruhe, Germany) using a 12 wt % KCl solution. The KCl solution was broken into fine droplets by compressed air in the aerosol generator. The droplets were dried using a silica gel desiccator in the diffusion dryer, and they became solid particles. Then, the particles were neutralized using ion stream from a high-voltage neutralizer (model: 3088, TSI Inc., Shoreview, MN, USA), in order to achieve the Boltzman charge equilibrium. After neutralization, the KCl particles were ready for the filtration test. All of the compressed air used in this system was filtered by a high efficiency particulate air (HEPA) filter, in order to obtain clean air (see Figure 1).

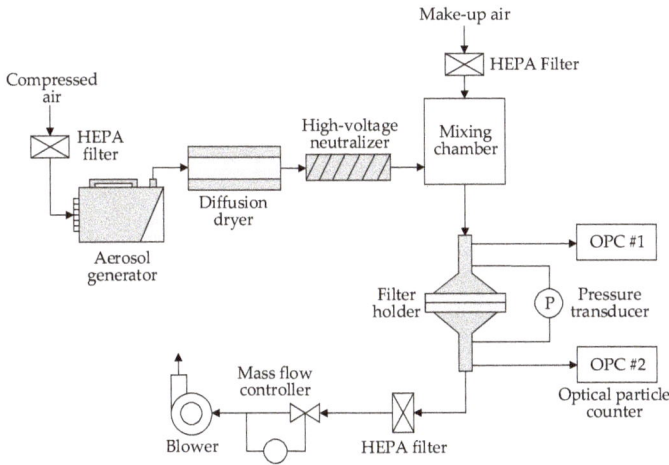

Figure 1. Schematic of filtration test setup.

The filter paper was mounted in the filter holder with the test area of 100 cm^2. The upstream and downstream of the filter holder were connected to a pressure transducer (Model: 166, Alpha Instruments Inc., Acton, MA, USA) and two optical particle counters (OPC, model: 3330, TSI Inc., Shoreview, MN, USA). The pressure transducer measured the pressure drop across the filter paper under a certain testing velocity. The optical particle counters sampled the air constantly from the test flow and measured the particle number concentration of the upstream and downstream air. The measurement was based on ISO/TS 19713-1:2010 [45]. The upstream number concentration, $N_{up,i}$ (#/cm^3), and downstream concentration, $N_{down,i}$ (#/cm^3), of the particles with a diameter of d_i were used to calculate the filtration efficiency, E_i, as shown in Equation (3). To obtain reliable results, the particle concentration should be appropriate to meet the minimum upstream counts ($\geq$500) and should avoid the effect of the particle deposition on the filtration performance. The particle mass concentration in this study is 10 mg/m^3, under which the pressure drop would not change for 30 minutes.

$$E_i = 1 - \frac{N_{down,i}}{N_{up,i}} \tag{3}$$

A good filter paper is one that gives a high collection efficiency with a low pressure drop. A useful criterion for comparing filter papers is the figure of merit [44], FOM_i (Pa^{-1}), of particles with a diameter of d_i, as shown in Equation (4).

$$FOM_i = \frac{\ln(1 - E_i)}{\Delta p} \tag{4}$$

where Δp is the pressure drop, Pa.

A mass flowmeter (Model: 4043, TSI Inc., Shoreview, MN, USA) and a ball valve on the downstream were used to control the test flow rate. The face velocity across the filter paper was 10.0 cm/s, which was frequently used in the air filter test for the engine intake system. The temperature of the ambient air was 21 °C with a relative humidity less than 10%.

3. Results and Discussion

3.1. Effect of Refining Pressure on Fibrillation of Lyocell Fiber

The microscopic images of the fibrillated Lyocell fiber under different refining pressure and revolutions can be seen in Figures 2 and 3. The magnification ratio (100:1) for each image was the same, which meant that the fiber size in the microscopic images was comparable. As can be seen, when the refining pressure was standard, 3.33 N/mm, the Lyocell fiber could be effectively fibrillated. When the number of revolutions reached 10,000, the fine nanofibers were fibrillated from the bulk Lyocell fiber. However, when the number of revolutions reached 40,000, the Lyocell fibers were cut into short fibers, and were compared with Figure 2a. It demonstrated that under a refining pressure of 3.33 N/mm, the energy from the refiner was largely used to cut the Lyocell fiber instead of the fiber fibrillation. The massive fragmented fibers indicated that the fibrillation under this refining pressure was unsatisfactory.

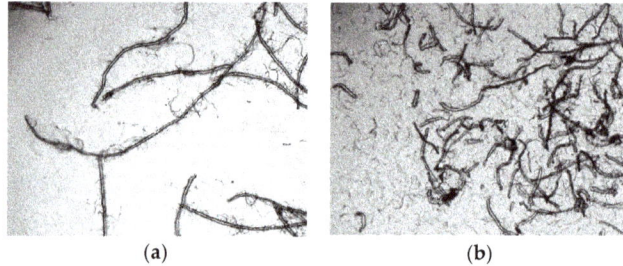

(a)   (b)

Figure 2. Microscopic images (×100) of fibrillated the Lyocell fiber under refining pressure of 3.33 N/mm: (**a**) 10,000 revolutions; (**b**) 40,000 revolutions.

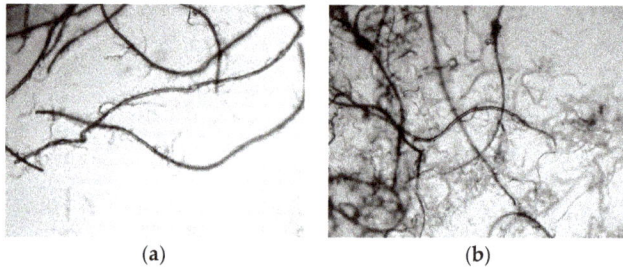

(a)   (b)

Figure 3. Microscopic images (×100) of the fibrillated Lyocell fiber under refining pressure of 1.00 N/mm: (**a**) 10,000 revolutions; (**b**) 40,000 revolutions.

When the refining pressure was 1.00 N/mm, the fiber fibrillation was more satisfactory. Although the fibrillation process was slower under the refining pressure of 1.00 N/mm, as can be seen in Figures 2a and 3a, the Lyocell fiber was less likely to be cut into short fibers. When the number of revolutions reached 40,000, abundant nanofibers were produced, but the Lyocell fibers were still long. The energy from the refiner was largely used to fibrillate the Lyocell fiber instead of cutting the fiber. As the energy consumption of the PFI refiner was related to the number of revolutions, it was apparent that a refining pressure of 1.00 N/mm was more efficient for the Lyocell fiber fibrillation. Based on the microscopic observation, the preferred refining pressure for the Lyocell fiber was 1.00 N/mm, which was selected for following studies.

3.2. Effect of Number of Revolutions on Fibrillation of Lyocell Fiber

The SEM images of the fibrillated fiber under a refining pressure of 1.00 N/mm can be seen in Figure 4, which demonstrated the fibrillation process. As shown in Figure 4a, the original Lyocell fiber was cylindrical, and the surface was very smooth. According to other studies [46], the Lyocell fiber had the island-in-the-sea structure and was made up of macrofibril (0.5–1 μm), while the macrofibril was made up of microfibril with diameter in the order of 100 nm. The Lyocell fiber with a skin-core structure had an amorphous skin of 35–80 nm [47]. In the initial stage of fibrillation, the skin of the Lyocell fiber was peeled off by the mechanical force of the PFI refiner, while the core of the Lyocell fiber was less affected by refining, as shown in Figure 4b. When the number of revolutions reached 10,000, most of the skin of the Lyocell fiber was peeled off, and individual or bundled macrofibrils were peeled off along the length of the Lyocell fiber, as can be seen in Figure 4c. When the number of revolutions reached 20,000, as shown in Figure 4d, more macrofibrils were peeled off, and the bundled macrofibrils were gradually split into individual macrofibrils. As the refining continued, most of the bundled macrofibrils were split into individual ones when the number of revolutions reached 40,000. The small amount of ultrafine fibers in Figure 4e indicated that small portions of macrofibrils were also split into microfibrils. The discussion of the fibrillation process showed that the cellulose nanofibers obtained under 1.00 N/mm and 40,000 revolutions were mainly the macrofibrils of the Lyocell fiber, and these nanofibers exhibited a range of fiber diameters.

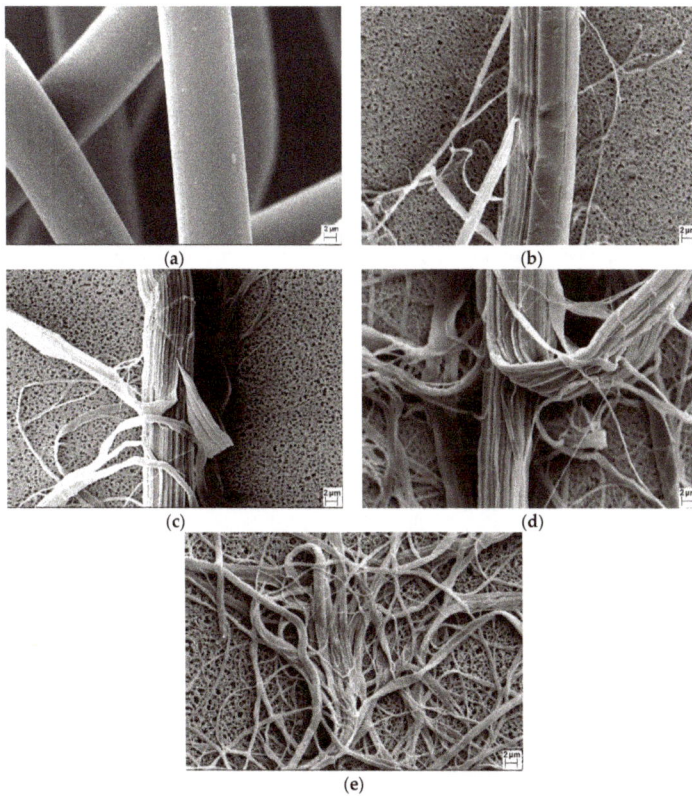

Figure 4. SEM images of fibrillated Lyocell fiber under a different number of revolutions: (**a**) original Lyocell fiber; (**b**) 5000 revolutions; (**c**) 10,000 revolutions; (**d**) 20,000 revolutions; (**e**) 40,000 revolutions.

The detailed properties of fibrillated fibers under different revolutions can be seen in Table 2. The beating degree increased from 13°SR to 79°SR when the number of revolutions increased. As the beating degree indicated the average fiber diameter, the tread of the beating degree agreed well with the changes of the fiber morphology in Figure 4. When the number of revolutions increased from 0 to 5000, the change of the beating degree was very small, as the fiber skin was peeled off but the core of the Lyocell fiber was less affected in this refining stage. When the number of revolutions was 10,000, the beating degree increased significantly as a result of the peeling of the macrofibrils from the Lyocell fiber. When 20,000 revolutions were reached, there was a sharp increase of the beating degree because of more individual macrofibrils. Under 40,000 revolutions, the beating degree continued to increase sharply, as most of the Lyocell fibers had been split into the fine macrofibrils.

Table 2. Properties of fibrillated fibers under different revolutions.

Number of Revolutions	Beating Degree, °SR	Average Fiber Length, mm	Fine Fibers Content, %	Specific Surface Area, m²/g	Average Fiber Diameter d_f, µm
0	13	4.00 ± 0.01	0	-	-
5000	18	3.16 ± 0.06	15.0 ± 0.2	-	-
10,000	31	2.66 ± 0.08	29.8 ± 1.1	-	-
20,000	58	1.47 ± 0.11	38.3 ± 1.2	2.42 ± 0.04	1.1 ± 0.09
40,000	79	1.31 ± 0.09	43.7 ± 1.5	3.33 ± 0.02	0.799 ± 0.06

It can also be found in Table 2 that the average fiber length decreased and the fine fibers content increased with the higher number of revolutions. It was noticeable that when the number of revolutions increased from 20,000 to 40,000, although the beating degree increased sharply, the changes of the average fiber length and the fine fiber content were small, which indicated that the refining energy was mostly used to spilt the bundled macrofibrils into finer fibrils in this refining stage. Because of the high longitudinal orientation of the crystallites of the Lyocell fiber [28–30], the average fiber length of the fibrillated nanofiber after 40,000 revolutions was more than 1 mm, which was much larger than the natural fiber under the same fibrillation condition. The specific surface area in Table 2 was used to calculate the average fiber diameter of the fibrillated fibers. As the fibrillated fibers from 0 to 10,000 revolutions owned a small specific surface area (smaller than 1.0 m²/g), it was very difficult to obtained accurate and reproducible results using the BET method. Only the specific surface area of the fibers under 20,000 and 40,000 are shown in Table 2. As can be seen, the average fiber diameter of the nanofibers under 40,000 were close to the diameter of the Lyocell microfibril (0.5–1 µm). It demonstrated that the cellulose nanofibers that were obtained under 40,000 revolutions were mainly macrofibrils of the Lyocell fiber, which agreed with the observation of the SEM images in Figure 4. As the Lyocell fiber was fully fibrillated with a good fiber length under 1.00 N/mm and 40,000 revolutions, the cellulose nanofibers under this condition were used to prepare the filter paper for the following tests.

3.3. Physical Properties of Filter Paper Containing Fibrillated Nanofiber

The physical properties of the filter paper containing different amounts of nanofibers are shown in Table 3. As can be seen, with the higher fraction of nanofibers, both the thickness and porosity slightly decreased, which showed that the cellulose nanofiber had little effect on the thickness and porosity of the filter paper. The mean pore size decreased dramatically with only a 5% nanofiber. Therefore, the nanofiber can reduce the pore size of the filter paper very efficiently. The mean pore size for the filter with a 20% nanofiber was 5.0 µm, which was larger than the $PM_{2.5}$. It indicated that sieving was not a main filtration mechanism in this work, and that the particles were supposed to be captured by coarse fibers and nanofibers due to impaction, direct interception, and diffusion.

Table 3. Physical properties of filter paper containing nanofiber.

Filter Paper No.	Weight Fraction of Nanofibers	Basic Weight, g/m^2	Thickness, mm	Porosity	Mean Pore Size, µm
#1	0	104.3 ± 1.0	0.452 ± 0.012	0.843 ± 0.020	22.6 ± 0.5
#2	5%	104 ± 1.5	0.408 ± 0.010	0.827 ± 0.013	13.5 ± 0.3
#3	10%	103.6 ± 1.3	0.415 ± 0.009	0.830 ± 0.023	10.2 ± 0.3
#4	15%	103.6 ± 1.6	0.400 ± 0.010	0.824 ± 0.031	6.1 ± 0.2
#5	20%	104 ± 0.9	0.390 ± 0.011	0.819 ± 0.021	5.0 ± 0.2

The fiber microstructure of the filter paper containing different amounts of nanofibers are shown in Figure 5. For the filter paper without nanofibers, the fibrous structure was very open. With the presence of nanofibers in the filter paper, the fibrous structure greatly changed. As can be observed from Figure 5b to Figure 5e, there were two main ways of nanofiber arrangement in the filter paper, entangling with a coarse fiber or bridging the pores among the coarse fibers. When the nanofibers were entangled with the coarse fibers, the nanofibers would not contribute significantly to the filtration efficiency as expected. When the nanofiber bridged the pores, it would take advantage of the fine fiber size and lead to a good filtration performance. Therefore, from the perspective of filtration, it was preferred that the nanofibers were not entangled with other fibers. The impact of the nanofiber structure on the filtration performance is discussed in the next section.

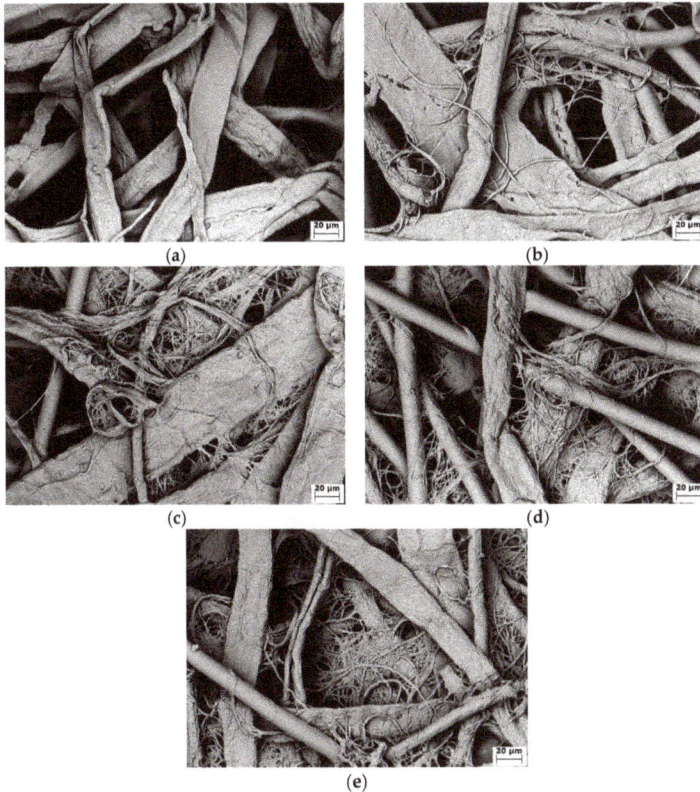

(a)

(b)

(c)

(d)

(e)

Figure 5. SEM images of filter paper containing different amounts of nanofibers: (**a**) sample #1; (**b**) sample #2; (**c**) sample #3; (**d**) sample #4; (**e**) sample #5.

3.4. Filtration Performance of Filter Paper Containing Fibrillated Nanofiber

The pressure drop and fractional efficiency are critical parameters for the filter paper. The pressure drop of the filter paper with a different fraction of nanofibers is shown in Figure 6. The pressure drop increased with the higher fractions of nanofibers. When the nanofiber fraction reached 15%, the pressure drop increased sharply. The fractional efficiency of the different filter papers can be seen in Figure 7. The efficiency against the particles ranging from 30 nm to 2 μm increased gradually with the higher nanofiber fraction. The particle size at which the efficiency was lowest in the efficiency curve was the most penetration particle size (MPPS). The MPPS for all five of the cases was about 200 nm. The efficiency at the MPPS increased from 0.1 to 0.6, which indicated that the cellulose nanofiber made by the Lyocell fiber was effective in improving the filtration efficiency of the filter paper.

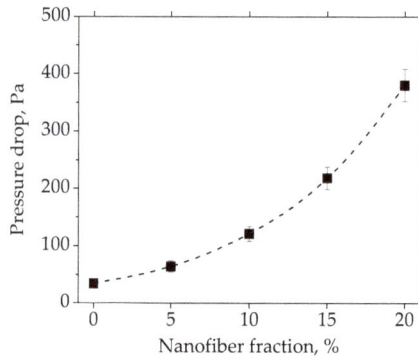

Figure 6. Pressure drop of filter paper with different fractions of nanofibers.

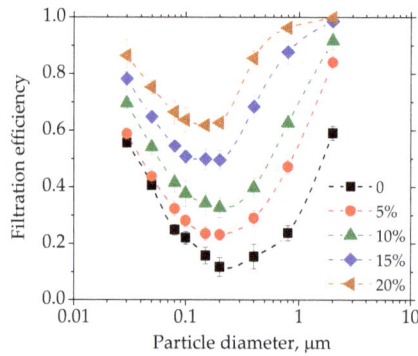

Figure 7. Fractional efficiency of filter paper with different fractions of nanofibers.

To clarify the effects of the nanofiber fractions on the filtration performance, the figure of merit was used. The figure of merit was defined as the ratio of the efficiency per unit of thickness to the pressure drop per unit thickness [48], which meant that the only factor affecting the figure of merit was the difference of the fibers. The calculated results of the figure of merit of the filter paper with a different nanofiber fraction against the particles ranging from 0.03 μm to 2 μm, are shown in Figure 8. For the particles smaller than 0.05 μm, the nanofiber had an adverse effect on the figure of merit. This result agreed with the study of Jing et al. [11], whose theoretical calculations showed that the figure of merit decreased as the fiber size decreased for the very small particles (below 50 nm). It was because the dominant filtration mechanism for the particles smaller than 0.05 μm was the Brownian diffusion [5,49], and the coarse fiber had a better performance in the diffusion regime. For the particles larger than 0.05 μm, the figure of merit increased substantially when only 5% of the nanofiber was used, as the filtration mechanism of interception and impaction [5,49], on which the fiber size had a significant effect, became more important. However, when the nanofiber fraction reached 10% and higher, the figure of merit began to decrease gradually. As the particles larger than 0.5 μm contributed to a major part of the PM$_{2.5}$ [43], the optimal fraction of the cellulose nanofiber against the particles in the PM$_{2.5}$ regime was 5%. The results of the figure of merit were related to the inhomogeneous distribution of the nanofibers in the fibrous structure, based on the SEM images of Figure 5. As mentioned above, the nanofiber that bridges the pores among the coarse fibers was better than the entangling with coarse fibers. When the nanofiber fraction increased, it was more likely that the nanofibers formed fiber clusters and were entangled with the coarse fibers. Therefore, the proper way to use the cellulose fiber in the wet-laid paper was to keep the nanofiber fraction low.

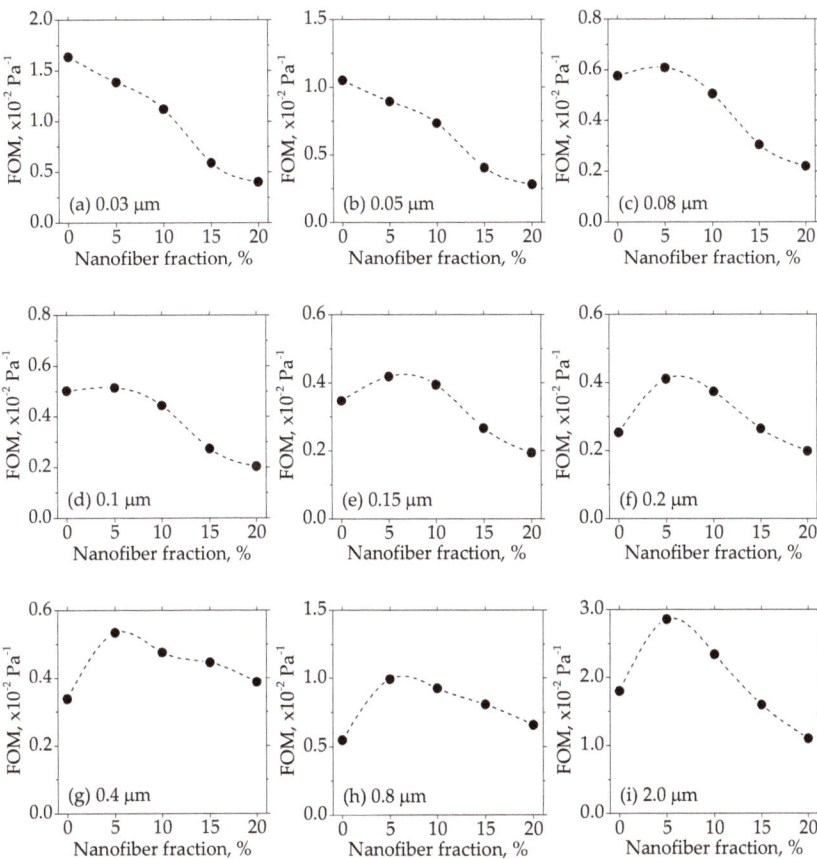

Figure 8. Figure of merit (FOM) of the filter paper against particles with different diameters.

3.5. Theoretical Analysis of Nanofiber Inhomogeneity

To characterize the inhomogeneity of the nanofibers in filter paper, Kirsch et al. [50] introduced the discrepancy between the theoretical and experimental pressure drop as the indicator of fiber inhomogeneity. For the theoretical calculation, the pressure drop across a filter paper can be calculated as the sum of the drags on all of the fibers. As the nanofibers were used, a gas slip effect should be considered. Pich [51] proposed Equation (5) to calculate pressure drop (Δp, Pa), by using the Kuwabara field to account for the slip effect at the gas-fiber interface.

$$\Delta p = \frac{16\mu l U_0 \alpha (1 + 1.996 Kn)}{d_f^2 [Ku + 1.996 Kn(-0.5\alpha - 0.25 + 0.25\alpha^2)]} \tag{5}$$

where μ is the air viscosity, Pa·s; l is the thickness of the filter paper, m; U_0 is the air face velocity, m/s; α is the fiber packing density ($\alpha = 1 - \varepsilon$); and d_f is the average fiber diameter, m. Kn and Ku are Knudson number and Kuwabara number, and they can be calculated by Equations (6) and (7), respectively.

$$Kn = \frac{2\lambda}{d_f} \tag{6}$$

$$Ku = -0.5 \ln \alpha - 0.75 + \alpha - 0.25\alpha^2 \tag{7}$$

where λ is the mean free path of air, m.

Equation (5) can be used for the filter paper with a uniform fiber diameter. The average fiber diameter for the nanofiber/microfiber mixture, however, was difficult to determine. In order to calculate the theoretical pressure drop of the multi-fiber paper, this study proposed the following method. The multi-fiber paper can be considered as the combination of several sub-papers made by the same fiber, as shown in Figure 9.

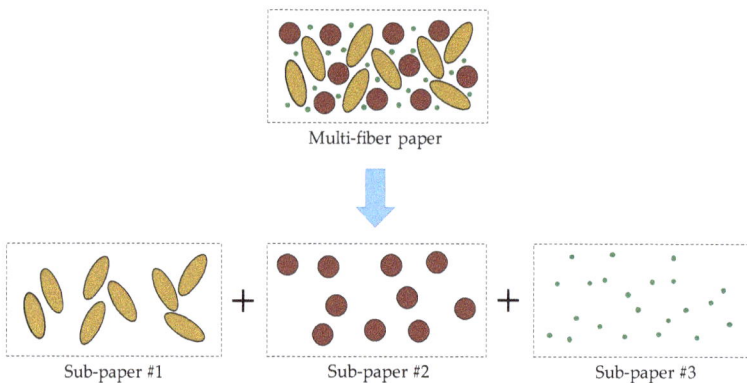

Figure 9. Schematic of multi-fiber paper and sub-papers.

The thickness of each sub-paper was the same as the multi-fiber filter paper. The fiber packing density of the sub-papers was determined by the volume fraction of each fiber in the multi-fiber filter paper, $f_{V,i}$, which can be calculated using Equation (8), as follows:

$$f_{V,i} = \frac{f_{m,i}/\rho_i}{\sum\limits_{k=1}^{3}(f_{m,k}/\rho_k)} \tag{8}$$

where the subscript i (i=1,2,3) refers to the softwood fiber, PET fiber, and cellulose nanofiber, respectively; $f_{m,i}$ is the weight fraction of each fiber in the multi-fiber filter paper, as shown in Table 1; and ρ_i is the density of each fiber, kg/m^3.

The fiber packing density of each fiber, which equals the fiber packing density of the corresponding sub-paper, can be calculated by Equation (9), as follows.

$$\alpha_i = f_{V,i}(1 - \varepsilon) \tag{9}$$

The theoretical pressure drop of the sub-paper can be calculated by applying Equation (9) into Equation (5). The overall pressure drop of the multi-fiber filter paper is the sum of the pressure drops of all of the sub-papers, as shown in Equation (10).

$$\Delta p = \Delta p_1 + \Delta p_2 + \Delta p_3 \tag{10}$$

The comparison of the theoretical and experimental results is shown in Figure 10. For the nanofiber fractions of 0 and 5%, the theoretical value was close to the experimental result. With the higher nanofiber fraction, the discrepancy between the theoretical and experimental results became larger. As the nanofibers were the major contributors of the pressure drop, they represented a higher degree of fiber inhomogeneity. As mentioned above in the fibrous structure analysis, the higher nanofiber fraction was more likely to result in fiber clusters and the entanglement with coarse fibers, which meant a lower nanofiber fiber inhomogeneity. The theoretical analysis agreed with the fibrous structure analysis, which confirmed that the optimal nanofiber fraction of 5% was due to the improved nanofiber homogeneity. As the filter papers in this study were prepared by standard procedures, some works will need to be done in the future in order to study the possibility of increasing the nanofiber homogeneity in the wet-laid preparation process, so as to maximize the filtration advantages of the nanofibers when the nanofiber fraction is higher.

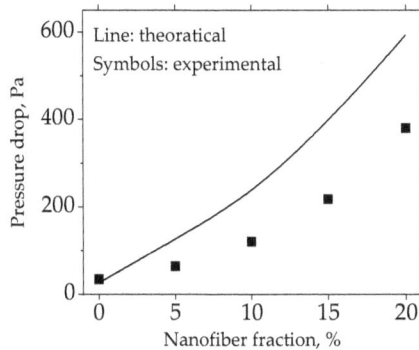

Figure 10. Experimental and theoretical pressure drop of filter paper.

4. Conclusions

In this work, cellulose nanofibers were prepared by Lyocell fiber nanofibrillation, and the influence of applying the cellulose nanofibers on filter paper was studied. It was found that the refining pressure of 1.00 N/mm was more satisfactory to fibrillate the Lyocell fiber, while the fibers were largely cut under the standard refining pressure of 3.33 N/mm. In the nanofibrillation process, the skin of the Lyocell fiber was peeled off first. Then, the individual or bundled macrofibrils were peeled off along the length of the Lyocell fiber. With further fibrillation, more macrofibrils were peeled off, and the bundled macrofibrils were gradually split into individual macrofibrils. The analysis of the fibrillation process showed that the cellulose nanofibers obtained under 1.00 N/mm and 40,000 revolutions were mainly macrofibrils of the Lyocell fiber with an average fiber diameter of 0.8 μm, which was suitable for filter paper. For the filter paper containing different fractions of cellulose nanofibers, both the pressure drop and fractional efficiency increased with a higher fraction of nanofibers. The results of the figure of merit demonstrated that for the particles smaller than 0.05 μm, the nanofiber had an adverse effect on the figure of merit, while for the particles larger than 0.05 μm, the figure of merit increased substantially with 5% nanofibers, but decreased when the nanofiber fraction reached 10% and higher. It was concluded that the optimal fraction of the cellulose nanofiber against the $PM_{2.5}$ was 5%. The results of the figure of merit were related to the inhomogeneous distribution of the nanofiber in the fibrous structure. Based on the SEM observation, when the nanofiber fraction increased, it was more likely that the nanofibers formed fiber clusters and entangled with the coarse fibers. The theoretical analysis of the nanofiber inhomogeneity indicated that the higher nanofiber fraction led to a higher degree of fiber inhomogeneity, which confirmed that the optimal nanofiber fraction of 5% was due to an improved nanofiber homogeneity.

Supplementary Materials: The following are available online at http://www.mdpi.com/1996-1944/11/8/1313/s1, Table S1: Technical data sheet of Lyocell fiber.

Author Contributions: Conceptualization, J.L. and M.T.; methodology, J.L.; software, M.T.; validation, J.L.; formal analysis, J.L. and M.T.; investigation, J.L.; resources, Y.L.; data curation, J.L.; writing (original draft preparation), J.L.; writing (review and editing), M.T. and Y.L.; supervision, J.H.; project administration, Y.L.; funding acquisition, J.H.

Funding: This research was funded by the Science and Technology Planning Project of Guangdong Province, Grant No. 2005B090925003, and the National Key R and D Program of China, Grant No. 2017YFB0308000.

Acknowledgments: The authors acknowledge Lenzing AG for providing the Lyocell fiber for this research.

Conflicts of Interest: The authors declare no conflict of interest.

References

1. Pui, D.Y.H.; Chen, S.C.; Zuo, Z. $PM_{2.5}$ in China: Measurements, sources, visibility and health effects, and mitigation. *Particuology* **2014**, *13*, 1–26. [CrossRef]
2. Gupta, A.; Novick, V.J.; Biswas, P.; Monson, P.R. Effect of humidity and particle hygroscopicity on the mass loading capacity of high efficiency particulate air (HEPA) filters. *Aerosol Sci. Technol.* **1993**, *19*, 94–107. [CrossRef]
3. Joubert, A.; Laborde, J.C.; Bouilloux, L.; Calle-Chazelet, S.; Thomas, D. Influence of humidity on clogging of flat and pleated HEPA filters. *Aerosol Sci. Technol.* **2010**, *44*, 1065–1076. [CrossRef]
4. Poon, W.S.; Liu, B.Y.H. A bimodal loading test for engine and general purpose air cleaning filters. *SAE Tech. Pap.* **1997**, *970674*. [CrossRef]
5. Wang, C.; Otani, Y. Removal of nanoparticles from gas streams by fibrous filters: A review. *Ind. Eng. Chem. Re.* **2012**, *52*, 5–17. [CrossRef]
6. Thomas, D.; Penicot, P.; Contal, P.; Leclerc, D.; Vendel, J. Clogging of fibrous filters by solid aerosol particles experimental and modelling study. *Chem. Eng. Sci.* **2001**, *56*, 3549–3561. [CrossRef]
7. Hutten, I.M. *Handbook of Nonwoven Filter Media*, 2nd ed.; Elsevier Butterworth Heinemann: Oxford, UK, 2012; pp. 520–523.
8. Chen, D.R.; Pui, D.Y.H.; Liu, B.Y.H. Optimization of pleated filter designs using a finite-element numerical model. *Aerosol Sci. Technol.* **1995**, *23*, 579–590. [CrossRef]
9. Wang, J.; Kim, S.C.; Pui, D.Y.H. Investigation of the figure of merit for filters with a single nanofiber layer on a substrate. *J. Aerosol Sci.* **2008**, *39*, 323–334. [CrossRef]
10. Podgórski, A.; Bałazy, A.; Gradoń, L. Application of nanofibers to improve the filtration efficiency of the most penetrating aerosol particles in fibrous filters. *Chem. Eng. Sci.* **2006**, *61*, 6804–6815. [CrossRef]
11. Wang, J.; Kim, S.C.; Pui, D.Y.H. Figure of merit of composite filters with micrometer and nanometer fibers. *Aerosol Sci. Technol.* **2008**, *42*, 722–728. [CrossRef]
12. Bao, L.; Seki, K.; Niinuma, H.; Otani, Y.; Balgis, R.; Ogi, T.; Gradon, L.; Okuyama, K. Verification of slip flow in nanofiber filter media through pressure drop measurement at low-pressure conditions. *Sep. Purif. Technol.* **2016**, *159*, 100–107. [CrossRef]
13. Leung, W.W.F.; Hung, C.H.; Yuen, P.T. Experimental investigation on continuous filtration of sub-micron aerosol by filter composed of dual-layers including a nanofiber layer. *Aerosol Sci. Technol.* **2009**, *43*, 1174–1183. [CrossRef]

14. Lavoine, N.; Desloges, I.; Dufresne, A.; Bras, J. Microfibrillated cellulose—its barrier properties and applications in cellulosic materials: A review. *Carbohydr. Polym.* **2012**, *90*, 735–764. [CrossRef] [PubMed]

15. Huang, P.; Zhao, Y.; Kuga, S.; Wu, M.; Huang, Y. A versatile method for producing functionalized cellulose nanofibers and their application. *Nanoscale* **2016**, *8*, 3753–3759. [CrossRef] [PubMed]

16. Fan, B.; Chen, S.; Yao, Q.; Sun, Q.; Jin, C. Fabrication of cellulose nanofiber/ALOOH aerogel for flame retardant and thermal insulation. *Materials* **2017**, *10*, 311. [CrossRef] [PubMed]

17. Ma, H.; Burger, C.; Hsiao, B.S.; Chu, B. Fabrication and characterization of cellulose nanofiber based thin-film nanofibrous composite membranes. *J. Memb. Sci.* **2014**, *454*, 272–282. [CrossRef]

18. Choo, K.; Ching, Y.C.; Chuah, C.H.; Julai, S.; Liou, N.S. Preparation and characterization of polyvinyl alcohol-chitosan composite films reinforced with cellulose nanofiber. *Materials* **2016**, *9*, 644. [CrossRef] [PubMed]

19. Abe, K.; Iwamoto, S.; Yano, H. Obtaining cellulose nanofibers with a uniform width of 15 nm from wood. *Biomacromolecules* **2007**, *8*, 3276–3278. [CrossRef] [PubMed]

20. Zhang, L.; Tsuzuki, T.; Wang, X. Preparation of cellulose nanofiber from softwood pulp by ball milling. *Cellulose* **2015**, *22*, 1729–1741. [CrossRef]

21. Jonoobi, M.; Mathew, A.P.; Oksman, K. Producing low-cost cellulose nanofiber from sludge as new source of raw materials. *Ind. Crops. Prod.* **2012**, *40*, 232–238. [CrossRef]

22. Halib, N.; Perrone, F.; Cemazar, M.; Dapas, B.; Farra, R.; Abrami, M.; Chiarappa, G.; Forte, G.; Zanconati, F.; Pozzato, G.; et al. Potential applications of nanocellulose—Containing materials in the biomedical field. *Materials* **2017**, *10*, 977. [CrossRef] [PubMed]

23. Iwamoto, S.; Lee, S.H.; Endo, T. Relationship between aspect ratio and suspension viscosity of wood cellulose nanofibers. *Polym. J.* **2014**, *46*, 73. [CrossRef]

24. Siró, I.; Plackett, D. Microfibrillated cellulose and new nanocomposite materials: A review. *Cellulose* **2010**, *17*, 459–494. [CrossRef]

25. Khalil, H.A.; Davoudpour, Y.; Islam, M.N.; Mustapha, A.; Sudesh, K.; Dungani, R.; Jawaid, M. Production and modification of nanofibrillated cellulose using various mechanical processes: A review. *Carbohydr. Polym.* **2014**, *99*, 649–665. [CrossRef] [PubMed]

26. Chakraborty, A.; Sain, M.; Kortschot, M. Cellulose microfibrils: A novel method of preparation using high shear refining and cryocrushing. *Holzforschung* **2005**, *59*, 102–107. [CrossRef]

27. Hubbe, M.A.; Rojas, O.J.; Lucia, L.A.; Sain, M. Cellulosic nanocomposites: A review. *BioResources* **2008**, *3*, 929–980.

28. Schurz, J.; Lenz, J. Investigations on the structure of regenerated cellulose fibers. *Macromol. Symp.* **1994**, *83*, 273–389. [CrossRef]

29. Crawshaw, J.; Bras, W.; Mant, G.R.; Cameron, R.E. Simultaneous SAXS and WAXS investigations of changes in native cellulose fiber microstructure on swelling in aqueous sodium hydroxide. *J. Appl. Polym. Sci.* **2002**, *83*, 1209–1218. [CrossRef]

30. Crawshaw, J.; Cameron, R.E. A small angle X-ray scattering study of the pore structure in tencel cellulose fibres and effects on physical treatments. *Polymer* **2000**, *41*, 4691–4698. [CrossRef]

31. Shibata, M.; Oyamada, S.; Kobayashi, S.I.; Yaginuma, D. Mechanical properties and biodegradability of green composites based on biodegradable polyesters and Lyocell fabric. *J. Appl. Polym. Sci.* **2004**, *92*, 3857–3863. [CrossRef]

32. Lenz, J.; Schurz, J.; Wrentschur, E. Properties and structure of solvent-spun and viscose-type fibres in the swollen state. *Colloid Polym. Sci.* **1993**, *271*, 460–468. [CrossRef]

33. Zhang, W.; Okubayashi, S.; Bechtold, T. Fibrillation tendency of cellulosic fibers—Part 3. effects of alkali pretreatment of Lyocell fiber. *Carbohydr. Polym.* **2005**, *59*, 173–179. [CrossRef]

34. Öztürk, H.B.; Okubayashi, S.; Bechtold, T. Splitting tendency of cellulosic fibers—Part 1. the effect of shear force on mechanical stability of swollen Lyocell fibers. *Cellulose* **2006**, *13*, 393–402. [CrossRef]

35. Yasumura, P.K.; DAlmeida, M.L.O.; Park, S.W. Multivariate statistical evaluation of physical properties of pulps refined in a PFI mill. *O Papel.* **2012**, *73*, 59–65.

36. Tang, M.; Hu, J.; Liang, Y.; Pui, D.Y.H. Pressure drop, penetration and quality factor of filter paper containing nanofibers. *Text. Res. J.* **2017**, *87*, 498–508. [CrossRef]

37. Choi, H.J.; Kumita, M.; Hayashi, S.; Yuasa, H.; Kamiyama, M.; Seto, T.; Tsai, C.J.; Otani, Y. Filtration properties of nanofiber/microfiber mixed filter and prediction of its performance. *Aerosol Air Qual. Res.* **2017**, *17*, 1052–1062. [CrossRef]

38. *Laboratory Beating of Pulp (PFI Mill Method), TAPPI T248 sp-15*; TAPPI Press: Atlanta, GA, USA, 2015.

39. *Pulps-Determination of Drainability-Part 1: Schopper-Riegler Method, ISO 5267-1*; International Organization for Standardization: Geneva, Switzerland, 1999.

40. *Pulps-Preparation of Laboratory Sheets for Physical Testing-Part 1: Conventional Sheet-Former Method, ISO 5269-1*; International Organization for Standardization: Geneva, Switzerland, 2005.

41. *Thickness (Caliper) of Paper, Paperboard, and Combined Board, TAPPI T411 om-15*; TAPPI Press: Atlanta, GA, USA, 2015.

42. *Standard Test Methods for Pore Size Characteristics of Membrane Filters by Bubble Point and Mean Flow Pore Test, ASTM F316-03*; ASTM International: West Conshohocken, PA, USA, 2011.

43. Thielke, J.F.; Charlson, R.J.; Winter, J.W.; Ahlquist, N.C.; Whitby, K.T.; Husar, R.B.; Liu, B.Y.H. Multiwavelength nephelometer measurements in los angeles smog aerosols. ii. correlation with size distributions, volume concentrations. *J. Colloid Interface Sci.* **1972**, *39*, 252–259. [CrossRef]

44. Tang, M.; Chen, S.C.; Chang, D.Q.; Xie, X.; Sun, J.; Pui, D.Y.H. Filtration efficiency and loading characteristics of $PM_{2.5}$ through composite filter media consisting of commercial HVAC electret media and nanofiber layer. *Sep. Purif. Technol.* **2017**, *198*, 137–145. [CrossRef]

45. *Road Vehicles-Inlet Air Cleaning Equipment for Internal Combustion Engines and Compressors-Part 1: Fractional Efficiency Testing with Fine Particles (0.3 µm to 5 µm Optical Diameter), ISO/TS 19713-1*; International Organization for Standardization: Geneva, Switzerland, 2010.

46. Schuster, K.C.; Aldred, P.; Villa, M.; Baron, M.; Loidl, R.; Biganska, O.; Patlazhan, S.; Navard, P.; Rüf, H.; Jerich, E. Characterising the emerging Lyocell fibres structures by ultra small angle neutron scattering (USANS). *Lenzinger Berichte.* **2003**, *82*, 107–117.

47. Ducos, F.; Biganska, O.; Schuster, K.C.; Navard, P. Influence of the Lyocell fibers structure on their fibrillation. *Cellul. Chem. Technol.* **2006**, *40*, 299–311.

48. Hinds, W. *Aerosol Technology: Properties, Behavior, and Measurement of Airborne Particles*, 2nd ed.; John Wiley & Sons: Hoboken, NJ, USA, 2012; pp. 187–188.

49. Brown, R.C. *Air Filtration: An Integrated Approach to the Theory and Applications of Fibrous Filters*; Pergamon Press: Oxford, UK, 1993; pp. 73–116.

50. Kirsch, A.A.; Stechkina, I.B. The theory of aerosol filtration with fibrous filters. In *Fundamentals of Aerosol Science*; Shaw, D.T., Ed.; John Wiley and Sons: Hoboken, NJ, USA, 1978; p. 165.

51. Pich, J. Pressure drop of fibrous filters at small Knudsen numbers. *Ann. Occup. Hyg.* **1966**, *9*, 23–27. [CrossRef] [PubMed]

materials

MDPI

Review

Phase Change Materials for Energy Efficiency in Buildings and Their Use in Mortars

Mariaenrica Frigione [1,*], Mariateresa Lettieri [2] and Antonella Sarcinella [1]

[1] Innovation Engineering Department, University of Salento, Prov.le Lecce-Monteroni, 73100 Lecce, Italy; antonella.sarcinella@unisalento.it

[2] Institute of Archaeological Heritage—Monuments and Sites, CNR–IBAM, Prov.le Lecce-Monteroni, 73100 Lecce, Italy; mariateresa.lettieri@cnr.it

* Correspondence: mariaenrica.frigione@unisalento.it; Tel.: +39-0832-297215

Received: 28 March 2019; Accepted: 15 April 2019; Published: 17 April 2019

Abstract: The construction industry is responsible for consuming large amounts of energy. The development of new materials with the purpose of increasing the thermal efficiency of buildings is, therefore, becoming, imperative. Thus, during the last decades, integration of Phase Change Materials (PCMs) into buildings has gained interest. Such materials can reduce the temperature variations, leading to an improvement in human comfort and decreasing at the same time the energy consumption of buildings, due to their capability to absorb and release energy from/in the environment. In the present paper, recent experimental studies dealing with mortars or concrete-containing PCMs, used as passive building systems, have been examined. This review is mainly aimed at providing information on the currently investigated materials and the employed methodologies for their manufacture, as well as at summarizing the results achieved so far on this subject.

Keywords: thermal energy storage (TES); phase change material (PCM); building materials; passive building systems; mortar; concrete

---

1. Introduction

The scientific community is severely concerned about the increase of world energy consumption. Global demand for energy is growing rapidly and higher consumption of fossil fuels leads to greater greenhouse gas emissions, particularly carbon dioxide (CO_2), which contribute to creating heavy environmental impacts, such as ozone layer depletion, global warming and climate change [1].

Among all the activities employing a great amount of energy, one of the main sectors in some countries is related to buildings. According to the International Energy Agency (IEA), the building sector accounts for more than 30% of total energy consumption (Figure 1a) and produces around 30% of total CO_2 emissions (even though when indirect building emissions from power generation are included, buildings and constructions represent nearly 40% of energy-related CO_2 emissions) (Figure 1b) [2].

The main cause for the intensified energy consumption is the overall change in the living standards and comfort demands for heating in cold regions and cooling in hot ones [3]. As a consequence, the energy efficiency of buildings is today a primary objective of policies regarding energy at regional, national and international levels [4]. The development of novel building materials able to improve the efficiency in energy utilization in the buildings is gaining increasing interest in industrial and academic communities. In addition, thermal energy storage (TES) is a useful tool for improving energy efficiency and increasing energy savings.

There are three ways to store thermal energy: chemical heat (CH, by breaking and forming molecular bonds), sensible heat (SH, by heating and cooling a material) and latent heat thermal energy storage (LHTES, by melting and solidifying a material) [5]. According to the literature, LHTES is the most attractive approach, due to its high storage capability and small temperature variations from storage to retrieval. In such a system, energy is stored during melting and recovered during freezing of a Phase Change Material (PCM) [6].

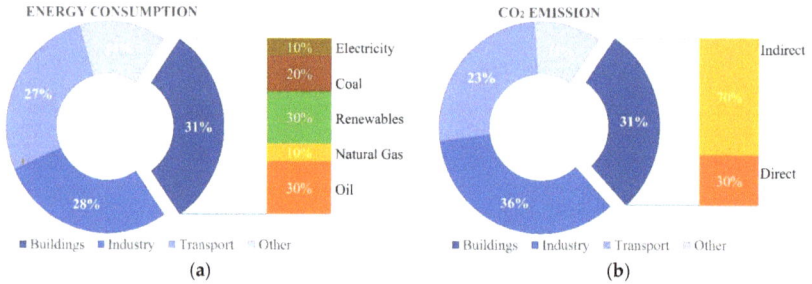

Figure 1. Energy consumptions in the building sector. (**a**) Energy consumption in each sector; (**b**) consequent CO_2 emissions.

Due to its capability to absorb and release energy from/in the environment, a PCM has the ability to reduce temperature variations. The operating principle of PCMs takes advantage of the modification of their state due to changes in temperature: as the temperature increases, the PCM passes from the solid to the liquid state, thus, absorbing and storing energy. Conversely, when the temperature decreases, the material can release the previously stored energy, passing from the liquid to the solid state, as illustrated in Figure 2 [7].

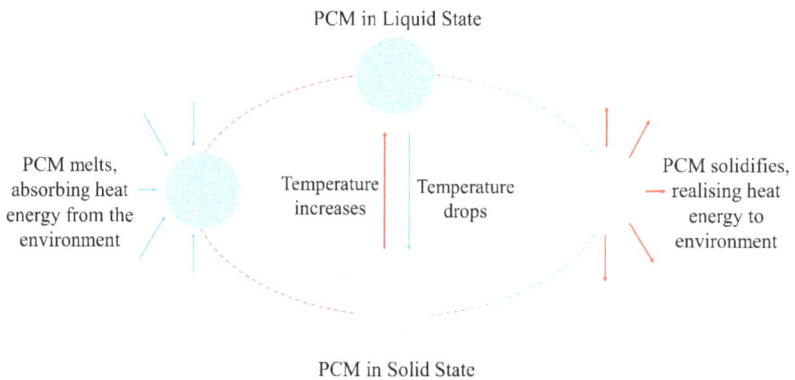

Figure 2. Phase change processes taking place in phase change materials (PCMs).

The incorporation in building materials of a suitable PCM can reduce the temperature fluctuations, thus, leading to an improvement in human comfort and a reduction in the consumption of energy in the building [8,9]. The use of PCMs in building materials is beneficial, especially in extremely hot and cold climates, where the energy required to maintain the internal conditions of buildings at a comfortable level can achieve significant consumption levels [10].

One of the oldest research on PCMs, describing the application of these materials in buildings, was published by Telkes [11] in 1975, followed by another work authored by Lane [12] in 1986. During the last decades, the integration of PCMs in building materials has gained renewed interest.

The PCMs in buildings are used in walls, floors, and ceilings, or in other building components (e.g., shutters and windows) as well as in heat and cold storage units [4,13]. Most of the applications in building structures consist of wallboards containing a PCM or in the incorporation of a PCM in a concrete or mortar matrix. The study of mortars developed using different binders with the incorporation of PCM largely interested the scientific community. The resulting properties of these new materials have been widely investigated, with a special focus on the thermal behavior in order to assess the advantages in terms of thermal energy storage [14].

This paper presents a literature review of the recent research works dealing with PCMs, taking as an example the production of mortars or concrete-containing phase change materials. After a general summary of the classification and properties of PCMs, the incorporation methods and the applications in the building sector are illustrated. The characteristics and features of different PCMs in mortar or concrete are, then, introduced, highlighting the main differences between the various mortars containing PCMs and describing the different methods employed for their production.

2. Classification of PCMs

The first classification of materials used for thermal storage appeared in 1983 and was proposed by Abhat [15]. Based on chemical composition, a PCM can be classified as an organic, inorganic or eutectic compound (Figure 3).

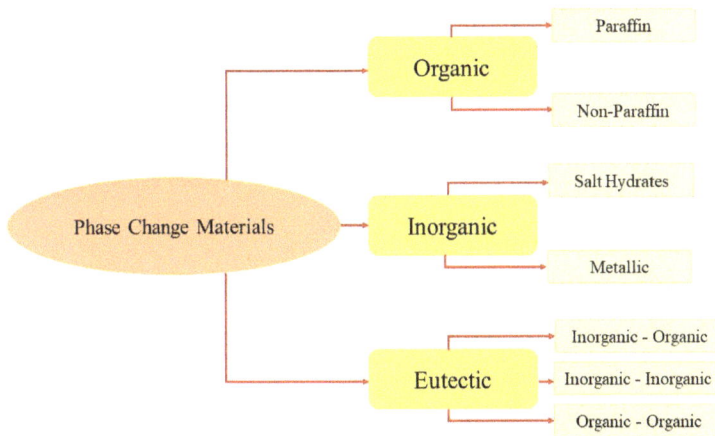

Figure 3. Classification of PCMs.

Organic materials, in turn, can be paraffinic or non-paraffinic. Typically, they can change their state several times without displaying any degradation. Salt hydrates and metals belong to a class of inorganic materials. The eutectic mixtures result from the combination of two or more organic and/or inorganic compounds, with the transition temperatures that can meet specific demands [16,17]. In Table 1, the main features of PCMs are summarized. The advantages and disadvantages of each class of PCMs are described in Table 2 [18,19].

Table 1. Main features of the different class of PCMs.

Type of PCM	Composition	Melting Temperature (°C)	Heat of Fusion (J/kg K)	Cost
Organic	Paraffin	−12–71	190–260	Costly
	Non-Paraffin	8–187	130–250	Highly costly
Inorganic	Salt hydrates	11–120	100–200	Low Cost
	Metallic	30–96	25–90	Costly
Eutectic	Paraffin	4–93	100–230	Costly
	Non-Paraffin	−12–71	190–260	Costly

Table 2. Main advantages and disadvantages of each type of PCMs.

Type of PCM	Advantages	Disadvantages
Organic	■ Available in a large temperature range; ■ Chemically inert; ■ Do not undergo phase segregation; ■ Thermally stable for repeated freeze/melt cycles; ■ Low vapor pressure in the melt form; ■ Relatively small melting heat; ■ Non-corrosive, or mildly corrosive (fatty acids); ■ Compatible with construction materials; ■ Small volume change during phase transitions; ■ Little or no super-cooling effect during freezing; ■ Innocuous (usually non-toxic and non-irritant; non-paraffin type may have various levels of toxicity); ■ Stable below 500 °C (non-paraffin type shows instability at high temperatures; ■ Recyclable.	■ Low thermal conductivity (around 0.2 W/m K); ■ Moderately flammable; ■ Non-compatible with plastic containers.
Inorganic	■ High volumetric storage heat; ■ High melting heat; ■ High thermal conductivity (0.5 W/m K); ■ Cheap and readily available; ■ Nonflammable; ■ Compatible with plastic containers; ■ Sharp phase change; ■ Low environmental impact; ■ Potentially recyclable.	■ Super-cooling during freezing; ■ Phase segregation during transitions; ■ Corrosive to metals; ■ Irritant; ■ High vapor pressure (inducing water loss and progressive changes in thermal behavior during thermal cycles); ■ Low durability (possible long term degradation when exposed to environmental agents); ■ Moderate chemical stability; ■ High volume change.
Eutectic	■ Sharp melting temperature; ■ High volumetric thermal storage capability (slightly lower than organic PCMs).	■ Limited data available on their thermo-physical properties.

It must be underlined, however, that not all existing PCMs can be used for thermal storage in building applications [20]. In order to be suitable for this use, at least two important requirements must be fulfilled: the PCM must display (i) appropriate melting temperature and (ii) melting heat.

The melting heat is a measure of the thermal energy that a material absorbs when it changes its state from solid to liquid. The thermal storage capacity of a PCM is strictly correlated to its melting heat [21].

Referring to the melting temperature, materials with a melting/freezing temperature between 18 °C and 40 °C are particularly suitable as PCM for building applications; this range of temperatures is, in fact, considered to be an optimum. According to the literature, the temperature of phase transition of the selected PCM should be very close to the human comfort temperatures (i.e., 22–26 °C) [22,23]. Nevertheless, PCMs that fall within three temperature ranges have been suggested for use in buildings [16]:

- Up to 21 °C for cooling applications;
- From 22 °C to 28 °C for optimal human comfort;
- From 29 °C to 60 °C for hot water applications, such as in the case of radiant floors often using water combined with PCMs.

The PCMs more suitable for building applications are summarized in Table 3, with the indication of the melting temperature and heat [8,13,14,19,24–34]. Only the materials with phase change temperatures ranging from 18 °C to 40 °C have been reported.

Table 3. PCMs for building applications: composition, category, melting temperature and melting heat.

PCM	Type	Melting Temperature (°C)	Melting Heat (J/kg K)
Glycerin	O	18	198.7
Hexadecane	O	18.1	236
$KF \cdot 4H_2O$	I	18.5	231
Butyl stearate	O	19	140
Propyl palmitate	O	19	186
Paraffin C_{16}–C_{18}	O	20–22	152
Heptadecane	O	20.8–21.7	171–172
Dimethyl sebacate	O	21	120–135
Octadecyl 3-mencaptopropylate	O	21	143
Lithium chloride ethanolate	O	21	188
$FeBr_3 \cdot 6H_2O$	I	21	105
Paraffin C_{17}	O	21.7	213
Erythritol palmitate	O	21.9	201
Polyglycol E600	O	22	127.2
Isopropyl stearate	O	22.1	113
Paraffin C_{13}–C_{24}	O	22–24	189
$34\%C_{14}H_{28}O_2 + 66\%C_{10}H_{20}O_2$	E	24	147.7
$50\%CaCl_2 + 50\%MgCl_2 \cdot 6H_2O$	E	25	95
Octadecane + docosane	E	25.5–27	203.8
$Mn(NO_3)_2 \cdot 6H_2O$	I	25.8	125.9
Octadecane + heneicosane	E	25.8-26	173.93
Octadecyl thioglycolate	O	26	90
Lactic acid	O	26	184
1-Dodecanol	O	26	200
$50\%CH_3CONH_2 + 50\%NH_2CONH_2$	E	27	163
Vinyl stearate	O	27–29	122
Paraffin C_{18}	O	28	244
Octadecane	O	28–28.1	244–250.7
Methyl palmitate	O	29	205

Table 3. *Cont.*

PCM	Type	Melting Temperature (°C)	Melting Heat (J/kg K)
$CaCl_2 \cdot 12H_2O$	I	29.8	174
$CaCl_2 \cdot 6H_2O$	I	29-30	171-192
$LiNO_3 \cdot 3H_2O$	I	30	296
Ga	I	30	80.9
$47\%Ca(NO_3)_2 \cdot 4H_2O + 53\%Mg(NO_3)_2 \cdot 6H_2O$	E	30	136
Capric acid	O	30.1	158
$60\%Na(CH_3COO) \cdot H_2O + 40\%CO(NH_2)_2$	E	30-31.5	200.5-226
Tridecanol	O	31.6	223
$Na_2SO_4 \cdot 10H_2O$	I	31-32.4	251.1-254
$Na_2SO_4 \cdot 3H_2O$	I	32	251
$Na_2CO_3 \cdot 10H_2O$	I	32-36	246.5-247
$CaBr_2 \cdot 6H_2O$	I	34	115.5
$LiBr_2 \cdot 2H_2O$	I	34	124
$Zn(NO_3)_2 \cdot 6H_2O$	I	35-36	265-281
$Na_2HPO_4 \cdot 12H_2O$	I	36-36.4	146.9-147
$FeCl_3 \cdot 6H_2O$	I	37	223
Tetradecanol	O	37.8	225
Camphenilone	O	39	205
Docasyl bromide	O	40	201
Caprylone	O	40	259

O = Organic; I = Inorganic; E = Eutectic.

3. Properties of PCMs

Among all possible candidates, the most appropriate PCM for a specific application must be selected taking into account some characteristics that will determine its effectiveness. For any thermal energy storage application in buildings, in fact, a careful examination of the overall properties of a PCM should be made, comparing the advantages and disadvantages displayed by each available system and, possibly, admitting a certain degree of compromise.

The main attractive properties and characteristics that a PCM shoud possess are reported in Figure 4 [16,19].

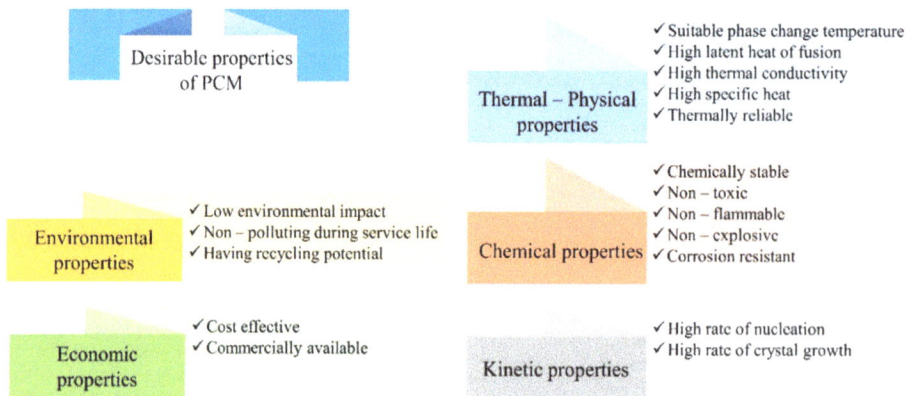

Figure 4. Advantageous properties of PCMs employed in building applications.

4. PCMs in Building Materials

Although this review mainly focuses on passive building systems for thermal energy storage based on the integration of PCM in building materials, a short overview of all the available solutions is presented.

Generally speaking, the possible introduction of PCMs in building materials is described as follows [4,5,34–36].

I.   Free cooling. This system requires a storage unit to accumulate the thermal energy and use it in heat absorption and in heat release. In this way, the storage medium is used to maintain a cold temperature, when the ambient temperature is lower than room temperature. This process is carried out during the night; the cold air flows through the storage unit, removes heat from the liquid PCM through an electrical fan; at this point, the PCM starts to solidify. When the room temperature rises above a comfortable level, the cold stored in PCM is released. Thus, the PCM absorbs heat from the air, starting the transformation from solid to liquid state [23,26,30,37,38].

II. Peak load shifting. This method is based on the use of PCMs that shift the peak energy request far from the peak hours of electrical demand; the peak load may be split throughout the day reducing the highest peaks [5,13,39]. The cooling/heating stored in off-peak hours is used during an on-peak load [40]. Peak cooling load reductions can range from 10 to 57% [4].

III. Active building systems. The storage capability of PCMs can be used in systems such as solar heat pump systems, heat recovery systems, and floor heating systems. An example of incorporating PCMs in an active system is radiant floors [5]. These systems consist of a lightweight piped radiant floor, where an integrated PCM layer is aimed at buffering internal gains during the summer season without affecting the winter warming capacity [22].

IV. Passive building systems. For passive applications, PCMs are integrated into building materials to increase their thermal mass. The incorporated PCM melts during the daytime and solidifies during the night: this process can warm the environment during the day.

4.1. Typical Applications

Among all potential applications of PCMs in buildings, the incorporation in construction materials (passive building system), aimed at modifying their thermal properties, has proven to be the most interesting. The combination of building materials with PCMs is an efficient way to increase the thermal energy storage capacity of construction elements. Thereby, wallboards, floors, roof, concrete and other parts are integrated with PCMs in order to improve the thermal performance of the building. The most common solution for implementing PCMs in buildings is the installation of PCM into the interior side of the building envelope. Thus, the use of suitable PCMs in the interiors of the construction allows to absorb and release heat in any room during a large part of the day. Several experimental investigations showed how this strategy positively affects indoor climate and energy use.

Wallboards or plasterboards are very suitable components for the incorporation of PCMs [3–5,19, 34,41,42]. These elements are cheap and widely used in building applications, especially in lightweight constructions, to reduce the internal air temperature fluctuations. The PCMs can be incorporated in the panel in different ways, as described below.

A PCM can also be added to conventional and alveolar bricks in constructions [3,42].

To insert a PCM in a house roof, a sequence of panels is usually used [3–5,13,34,42], each one containing the PCM, or a layer of mortar/concrete, with frustum holes filled with the PCM [3,4,42,43]. The PCM placed in the roof can absorb both the incoming solar energy and the thermal energy from the surroundings; hence, it reduces the internal temperature fluctuations.

The floor is another part of the building that offers a large surface and, thus, a great storage capability. As already described, the floor can often act as an active building system, but it can also be employed in passive ones. In some applications, in fact, PCMs have been included in the concrete layer placed under the floor; PCM panels have been also employed as an overlay to substitute the floor [4,5,13,39,41,42,44]. Advantageous effects are obtained by integrating PCMs in a floor, since a great amount of energy is usually lost from the floor, due to the heat transfer with the ground.

Emerging solutions are those in which PCMs are placed in windows and sunshades [3–5,45,46]. In such applications, a PCM must fill the glass, frame, layer, or any other hollow part, such as the cavity of the shutters. The main issue of this application is due to the lack of transparency of PCMs in both their liquid and solid states. Hence, the windows using such systems are blurry, with a reduced transmission of daylight and solar radiation.

Finally, PCMs in mortars or plasters are also considered for interior finishing on walls and ceilings in residential buildings [3–5,19,41,42,47].

4.2. Methods of Incorporation

Using one or more of the described elements in a house, a significant improvement in energy efficiency is achieved. Such an approach in construction allows to activate thermal inertia and heat storage capability of each room, reducing the internal temperature fluctuations and improving the level of indoor thermal comfort [48].

Different methods have been used to incorporate PCMs in building materials, such as:

- Direct incorporation;
- Immersion;
- Use of micro or macro encapsulated PCMs;
- Addition of shape-stabilized PCMs;
- Addition of form-stable PCM composites.

It must be emphasized that the terms shape-stabilized and form-stable PCMs have been often considered as synonymous; the two methods, on the other hand, have some distinct characteristics, as following described [19].

The first research published on PCMs largely focused on their direct incorporation and immersion. Direct incorporation is the simplest, practical and economical method: in this case, in fact, the PCM is directly mixed with the construction material [4,45]. The PCMs, in liquid or powdered form, are added to a mixture of materials (such as lime, gypsum, cement paste or concrete) during its production. The main advantage lies in the simplicity and inexpensiveness of the procedure, since no extra equipment is required. However, some problems due to the leakage of PCM, when it is in its melting state, can occur, possibly leading to low fire resistance of the impregnated materials and even causing incompatibility between the mixed materials [3,17].

Referring to the immersion method, porous construction materials are immersed in the melted PCM; thus, the porous materials absorb the product by capillary rise [22]. Once again, the mechanical and durability properties of the construction elements can be affected by leakage of PCMs and its incompatibility with the substrate. In particular, the leaked PCM in contact with the cementitious binder may interfere with the hydration reactions [19,27,44]; it may also cause the corrosion of reinforcing steel that, in turn, affects the service life of the concrete structure [49].

Some organic PCMs are not stable under an alkaline environment, which is typical of concrete, and they can easily react with calcium hydroxide; this can lead to a modification of the PCM, with a consequent decline in its properties. Furthermore, if a reaction with PCM occurs during the curing process of fresh concrete, the hydration of the cement may be delayed or interrupted, with a consequent reduction in its final strength. To avoid such unwanted effects, the Portland CEM I was identified as the most appropriate cement to be used with a PCM [50].

In order to reduce any interference with the building materials, a well-established approach is the encapsulation of PCMs in a suitable shell material [51].

Two encapsulation approaches are reported in literature for PCMs: micro-encapsulation (Figure 5a,b) and macro-encapsulation (Figure 5c) methods [16,52]. In order to guarantee a long-term stability of the whole system, the PCM and the container material should display no chemical interactions, irrespective of the encapsulation method [53].

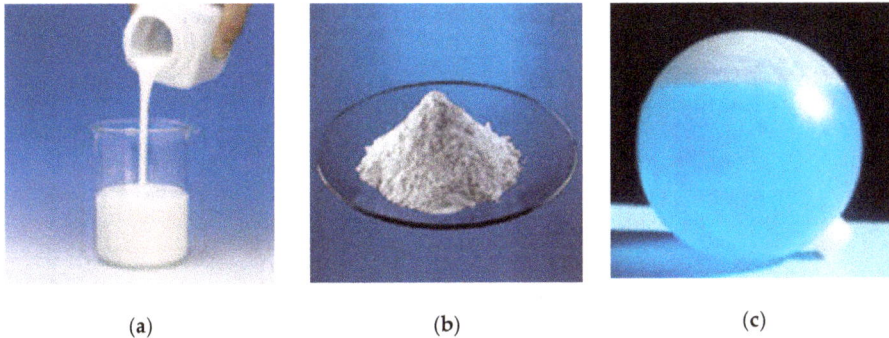

(a)   (b)   (c)

Figure 5. Examples of commercial PCMs: (a) micro-encapsulated PCM dispersed in a liquid; (b) micro-encapsulated PCM in powder form; (c) macro-encapsulated PCM in spherical form. Reprinted with permission from [5]. Elsevier (2015).

In the micro-encapsulation method, small PCM particles, ranging from 0.1 μm to 1 mm, are wrapped in a thin solid shell. The latter is usually constituted by natural or synthetic polymers [54]; in general, a shell of a high molecular-weight polymer is used. The employed polymer must be compatible with both the PCM and the construction material in which it is applied [50]. Different physical and chemical methods have been developed for the production of micro-encapsulated PCMs [3,35]. The physical methods include: pan coating, air-suspension coating, centrifugal extrusion, vibrational nozzle and spray drying. The chemical methods are: coacervation, complex coacervation and interfacial methods [13,14,27,32,33,52,53,55–59]. Chemical methods are likely to produce much smaller encapsulated PCM particles in comparison with the micro-encapsulated PCMs produced by physical methods [25].

The micro-encapsulation techniques present several advantages over the other procedures previously described [3,16]. These methods allow a reduction in leakage of PCM during its phase transition; they provide a high rate of heat transfer due to a large surface area per unit volume; they can improve chemical stability and thermal reliability (the latter representing the capability to repeat many times the melt/freeze cycle without the occurrence of degradation phenomena) [51]. Furthermore, the previous listed aspects contribute to expand the possibilities of integration of PCMs in construction materials [60]. Nonetheless, some issues still exist: the rigidity of the shell prevents natural convection, thus, reducing the rate of heat transfer [22,28]; micro-encapsulation may affect the mechanical properties of the building material [45]; finally, the very high costs limit this technique to high-value applications [61].

In the macro-encapsulation method, a significant amount of PCM is stored in rigid containers, such as tubes, spheres or panels, that are subsequently introduced in the construction elements. It is possible to produce, therefore, a system easier to move and handle and properly designed to satisfy the required application. The PCMs obtained with macro-encapsulation usually have poor thermal conductivity and the tendency to solidify at the edges. These materials need protection (for instance against drilling holes in the walls) and require that their introduction in the structure is performed in situ [3,19,60].

Among the different PCM encapsulating methods, the shape-stabilization one proved to be a very promising technique, although very complex in the implementation [35]. The shape-stabilized PCMs display several advantages, such as: large apparent specific heat, suitable thermal conductivity, the ability to maintain the shape during the phase-change process; moreover, they are thermally reliable over a longer service period, thus, the melt/freeze cycle responsible for the phase change behavior can be efficiently repeated many times. Amounts of PCM up to 80%, as a percentage of the total weight, can be used with this technique [3,13,17,62,63].

The shape-stabilized PCMs can be obtained by physical methods (such as blending, adsorption, and impregnation) or chemical methods (including graft copolymerization and sol-gel methods) [33]. The material supports used to fabricate shape-stabilized PCMs may consist of organic materials, based on polymers, such as high-density polyethylene (HDPE), styrene and butadiene, but also of inorganic porous materials [55]. The supports are usually able to prevent the leakage of the PCM [27]. The use of waste materials or by-products of different industrial processes may improve the sustainability of this technique [64,65].

In the case of polymers, the PCM and the supporting material are melted and mixed at high temperatures. Then, the polymeric support is cooled below its glass transition/melting temperature, solidifying, while the PCM, that is still in the liquid state, fills the empty space of the support (Figure 6).

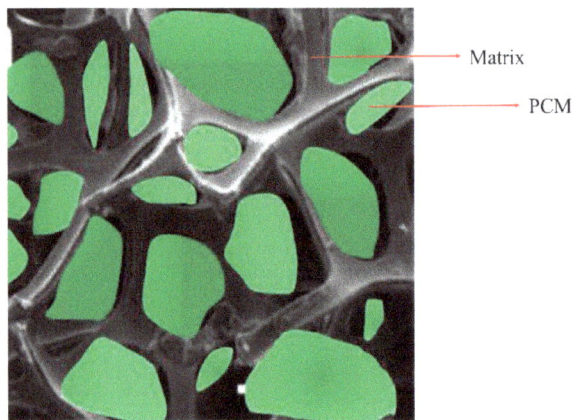

Figure 6. Example of shape-stabilized composite PCMs. Adapted with permission from [55]. Copyright 2015 Elsevier.

In order to improve the thermal conductivity of shape-stabilized PCMs, some additives can be added [19]. In particular, the shape-stabilized PCM composites show enhanced thermal conductivity with the incorporation of carbon-based nanostructures (CNs), namely: expanded graphite (EG), compressed expanded natural graphite (CENG), nano-graphite (NG), exfoliated graphite nanoplatelets (xGNPs), graphene, nitrogen-doped graphene (NDG), graphene oxide (GO) or multi-wall carbon nanotubes (MWCNTs).

A porous matrix of inorganic (such as silica-based material, perlite, diatomite, silica dioxide, clay materials) can act as the support containing the form-stabilized PCM [18,62,66–68]. In this case, the form-stable PCM composite can be obtained by immersing the matrix in the liquid PCM; a vacuum system can be employed to force the impregnation. Various vacuum impregnation systems have been used to prepare form-stable PCM composites [19]; an example is shown in Figure 7.

The term form-stable composite PCM is specifically used to define a composite material retaining an optimum/maximum percentage of PCM and showing no leakage when the temperature of the composite approaches the melting point of the PCM [69]. This feature also extends the durability of the material, preventing its degradation upon thermal cycling. This represents an important advantage for applications requiring long-term performance, like those related to buildings [63].

In Table 4, the advantages and disadvantages of each incorporation method are briefly described.

Materials **2019**, *12*, 1260

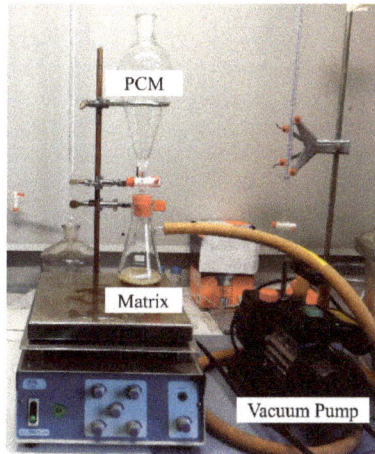

Figure 7. Vacuum impregnation equipment used to prepare form-stable PCM composites.

Table 4. Advantages and disadvantages of different methods for incorporation of PCMs into building materials.

Method of Incorporation	Advantages	Disadvantages
Direct incorporation	Simple and cheap	Possible leakage of PCM in the melting state; flammability of the impregnated elements is possible, as well as incompatibility between the materials.
Direct impregnation	Simple, practical and cheap	Leakage and incompatibility can occur, affecting the mechanical properties and durability of the construction elements.
Micro-encapsulated PCM	Reduced leakage of PCM during phase transition; higher heat transfer rate; improved chemical stability and thermal reliability.	The capsules are expensive; their rigidity may prevent natural convection and reduce the heat transfer rate; the mechanical properties of the construction materials may be affected.
Macro-encapsulated PCM	A significant quantity of PCM is packed in the container; easiness and suitability for any specific application.	Poor thermal conductivity and tendency to solidification at the edges; introduction in the structure must be carried out in situ.
Shape-stabilized PCM	Large apparent specific heat; suitable thermal conductivity; ability to maintain the shape of PCM during the phase-change; thermal reliability over a long period of time; reduced leakage phenomena.	Complex equipment is needed for their preparation; need to use additives to improve the thermal conductivity.
Form-stable composite	Very cheap; retaining of high amount of PCM without leakage above its melting point.	Complex equipment is needed for their preparation.

5. PCMs in Mortars: Potential and Issues

The incorporation of phase-change materials in mortars employed in the interiors of buildings appears the most attractive solution in an attempt to minimize the massive energetic consumption related to building conditioning. Such an approach allows the regulation of the temperature inside buildings through latent heat energy storage, using only solar energy as a resource, thus, reducing the need of heating/cooling equipment. Incorporation of PCMs in mortar and concrete can be an efficient method due to the large heat exchange area surfaces; furthermore, the final functional material can be adapted in a wide variety of shapes and sizes. Being mortar and concrete widely used as construction materials, such PCM composites can be employed in any practical application. In addition, quality control can be easily achieved in the produced materials.

In recent years, the research on mortars containing PCMs has been mainly focused on cement and gypsum compositions, due to the good mechanical and thermal properties of these binders. At first, cement-based mortars employed in masonry were taken into consideration and PCMs incorporated into these cementitious systems were examined, analyzing their capability to improve the energy efficiency of building envelopes. The results of these studies have shown that the incorporation of PCM appreciably decreased the mechanical properties of the composite, especially in the case of cement pastes [70–73]. For such a reason, the research has moved towards the investigation on mortars for interior and/or exterior coatings, where high values of mechanical strength are not required. More recently, other PCM-mortar systems based on different binders, such as aerial lime [73–87], hydraulic lime [71,88,89], and, in some cases, geopolymers [90–93], have been developed and studied [74,78,80,83,90,92].

Generally speaking, both the amount of heat stored as well as the thermal conductivity of the final material increased upon the addition of a PCM in a mortar [75,89–91,94–96]; the PCM reduces the fluctuations in temperature and the use of the composite material can efficiently improve the indoor temperature comfort [79,97–99]. The concurrent use of other additives (e.g., TiO_2 nanoparticles) allows to obtain multifunctional materials [95] to prevent the accumulation of dirt and growth of microorganisms on the surfaces; they can even degrade pollutants.

5.1. Effects of PCM's Type and Content

It has been found that the higher the PCM content, the better the heat storage capacity of the mortar [100,101]. Comparing the heat storage in samples with and without PCM incorporation, the PCM composite can reduce the heat effect in the heat storage process as well as enhance the heat effect in the heat release process.

Generally speaking, the addition of a PCM in a mortar formulation requires a larger amount of water to guarantee the workability of the fresh mixture [72,87,102,103]. On the other hand, PCMs added as spherical particles can assure an appropriate workability of host mortars over time [95], since this shape appreciably reduces the surface friction.

Among the different types of PCMs, organic compounds have been found to be the most promising candidates for applications in mortars due to their chemical stability, non-corrosive nature, reproducible melting and crystallization behavior, even after repeated thermal cycles [104,105]. Moreover, more than one material can be used in a mixture, to achieve the desired phase transition temperature [99,106].

In the examined literature, paraffin compounds [65,70,73,74,76–91,95,97,100,102,107–128] are the most used PCM in mortars. Furthermore, higher hydrocarbons (like octadecane [72,94,101,129–132], hexadecane [98], or heptadecane [71]), fatty acids or their mixtures [99,106,133,134], and polyethylene glycols [64,96,135] have been extensively examined for these applications. Their phase change temperature usually ranges between 18 °C and 36 °C, and the phase change enthalpy is between 100 and 260 J/g. Some of these compounds have a low thermal conductivity which limits heat transmission, this latter representing a significant drawback in thermal exchange applications. However, the addition of fillers with high thermal conductivity could solve this problem; for instance, expanded graphite is widely used to increase the thermal conductivity of the PCMs [99–101,136].

The incorporation of PCM has been mostly performed by their addition in microcapsule form; their composition, when detailed, is generally based on polymethylmethacrylate or melamine–formaldehyde [67,70–72,74,76,78–87,102,117,121]. Their application through the form- or shape-stabilization methods is increasing [64,96,123,126,127,130,132,133], with perlite [90,106,122,124,125,134] and graphite [94,98–101] mainly employed as supports.

Salt hydrates have been rarely applied in mortars [137]; in some rare case, the PCM has been added by direct incorporation; in addition, it has been found not appropriate in combination with a cement-based binder.

5.2. Influence on the Porosity and Mechanical Properties of the Mortar

Mortars have shown better performance using microcapsules containing a PCM, in amounts from 15 to 20 wt.% (by total mass), with an excellent compromise between mechanical strengths, shrinkage, thermal efficiency, and costs [76,78,82,86,91,113,117]. In such cases, a strong reduction of macroporosity, rather than a decrease of total porosity, leads to an improvement in mechanical strength [76]. Usually, a lower total open porosity and a higher number of smaller pores are obtained in a PCM-modified mortar in comparison to the reference material [87–89]. The incorporation of PCM, in fact, produces a filling effect that reduces the number of macropores [76,88]. In a few cases, even though the pore size increases upon the addition of a PCM, the compressive strength rises due to the crystallization of minerals (such as aragonite) able to enhance the resistance of the mortar [75]. Leakage of the PCM into the mortar could increase its porosity [110] with a consequent reduction in mechanical properties. Large amounts of PCM (i.e., higher than 20 wt.%) have been found not useful, due to the great amounts of water required in these mortar mixes [79] causing a high porosity [78]. On the other hand, it has been found that a high PCM content does not necessarily imply an increase in the latent heat transfer [127], proving that the internal porosity plays an important role in this process [65,76,82,83]. The presence of nanopores leads to a decrease in the heat transfer capability, even though the PCM content is close to the optimal value.

The mechanical strength of mortars containing PCMs is strongly related to the microstructural characteristics of the material, that are, in turn, influenced by the nature/dosage of the components and by the amount of added water [74,95]. However, the content of PCM, rather than the type of binder, is the key factor affecting the mechanical behavior of the final mortar. The added PCMs are likely to behave more like voids than as aggregates in mortars and concretes; therefore, they do not contribute to the compressive strength [72]. When mechanical properties lower than those displayed by pristine materials are obtained [73,100,120,127], this result is mainly ascribed to the increased required amount of water due to incorporation of PCM [71,81,95], rather than to the relatively poor interface compatibility [120]. A greater strength is generally obtained when a lower amount of PCM is used [100,112]. A reduction in mechanical properties can also be observed when the addition of PCMs produces a decrease in hydration heat and/or a delay in the hydration kinetics [73,120,126]. Even though the PCM does not take part in the hydration processes, it can sequester water, thus, hindering hydration of the mortar. These effects influence the curing and, as a consequence, the mechanical properties of mortars that, especially at early stages, are lower when a PCM is added [110]. Furthermore, by increasing the content of the PCM, a greater number of incorporated PCM particles can be broken during shearing failure (Figure 8); the so formed void space, previously occupied by the un-broken PCM, causes an increase in the porosity of the matrix [113] and, consequently, lower shear strength and stiffness can be measured [70,91,117].

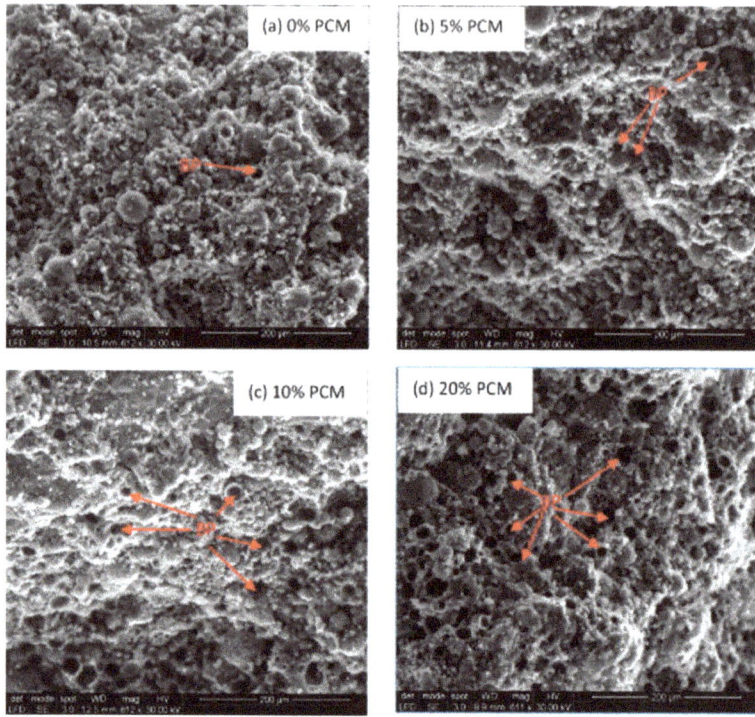

Figure 8. SEM image of failure surface of mortar specimens containing (**a**) 0% PCM, (**b**) 5% PCM, (**c**) 10% PCM, and (**d**) 20% PCM (BP = broken particles). Reprinted with permission from [91]. Elsevier (2015).

5.3. Main Issues: Shrinkage, Cracking, and Leaking

The addition of microcapsules of PCM significantly affects the shrinkage of the mortar [71,79]; the higher the amount of PCM, the higher the shrinkage. To avoid cracking, the shrinkage phenomena taking place during the curing process must be limited. To this aim, small amounts of PCM and a low water/binder ratio are usually recommended [95]. Some works also examined the capability of PCMs to mitigate early-age temperature rise in cementitious materials caused by exothermic cement hydration and the resultant risk of thermal cracking [138–140].

Damage of the PCM particles, taking place even during the initial mechanical mixing, may cause a lower heat capacity [73,119] and can also reduce the resistance of the mortar to fire [110].

The use of lightweight aggregates (LWAs) as a support for the PCM can reduce the interferences with the hydration reactions [123,126] as well as the leakage of PCM from the composite [122]. Furthermore, the addition of fiber and/or gypsum has been found to be a good solution to solve problems related to cracking caused by the incorporation of microcapsules in lime-based mortars [78–82]. The inclusion of stiff quartz can also significantly reduce shrinkage due to the aggregate restraint effects [119].

5.4. Analytical Characterization and Simulations by Prototypes

Besides the technical (composition, workability, density, porosity, thermal conductivity) and mechanical characterization, several tests and analytical techniques have been used to evaluate the properties and the behavior of the mortars containing PCMs. In particular, observations through a scanning electron microscope (SEM) have been carried out to control the microstructure of mortars [70, 71,76,80,83,113,115,133] and the impregnation of the support [90,130], or to reveal the damage of the encapsulated PCM particles [73,74,110,119,131,132]; this latter event frequently results in the leakage of the PCM.

The thermal characterization is usually carried out by differential scanning calorimetry (DSC). The shape of the DSC curves and the phase change temperatures calculated on the mortars containing PCM capsules and on pure PCM are similar [72,88], with the specific enthalpy proportional to the mass fraction of PCM in the mortar sample [114]. Conversely, the thermal properties of the PCM in a support, measured by DSC, are often found to be lower than those of pure PCM [64,90,96,124,132,134]. The difference observed between temperatures for melting and solidification, indicative of super-cooling, is common in PCMs [120] and it is ascribed to a lack of heterogeneous nucleation sites where the solidification is initiated [119,121]. DSC experiments performed at high heating/cooling rates intensify the differences between the heating and cooling processes, both in terms of the peak temperatures and the shape of the curves [102,114,115,122]. This hysteretic behavior in the thermal response of mortars with PCM is appreciably reduced as the heating/cooling rates are decreased [115].

Thermal analyses are performed to determine the effective content of PCM [94,98,120,130,132] or to evaluate the heat of hydration of the mortar [113]. The influence of PCMs on the hydration reactions and on the amounts of hydration products formed in these processes can be also evaluated by FT-IR (Fourier Transform Infrared Spectroscopy) analysis [70,123]. These latter investigations allow to examine the compatibility and interactions between the PCM and the matrix. Usually, the addition of PCM causes no new signals; only slight shifts of the characteristic absorption peaks of the components are observed. Such a result indicates that no chemical reactions, but only physical interactions, occur and that the integration of the PCM component takes place without a change in the chemical composition of the mortar [64,120].

Mortars containing blends of different micro-encapsulated PCMs, with different melting temperatures, have been investigated [115]. The final properties of such mortars were well predicted through the superposition of the effects of each PCM [114], as long as no interaction occurs between them.

Prototypes or numerical models have been often proposed to simulate and predict the behavior at the macro-scale level of mortars containing PCM [70,74,107–109,111,116,118,121,124,129], mainly in terms of thermo-physical properties of the final system containing PCM and relative cost implications. In most of these studies, the numerical solutions fitted very well with the corresponding experimental measurements, irrespective of the analyzed season. An example of a prototype used to experimentally evaluate the thermal performance of a mortar containing PCM is reported in Figure 9.

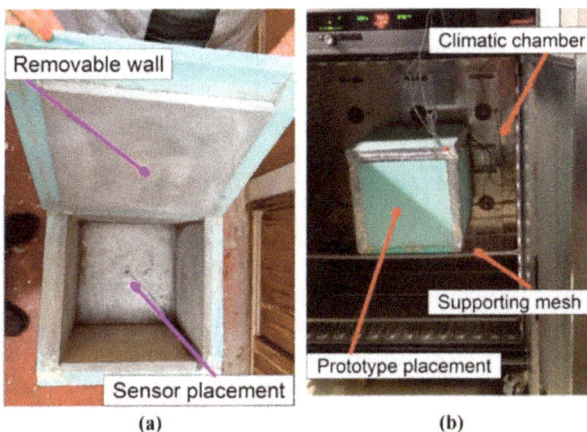

Figure 9. Prototype test cell including a PCM plastering mortar: (**a**) internal walls; (**b**) set-up within the climatic chamber for thermo-hygrometric cycles. Reprinted with permission from [116]. Elsevier (2015).

5.5. Durability

Very recently, different studies have been undertaken to assess the durability of mortars containing PCMs. Several aspects have been investigated; in particular, the structural integrity of the PCM during the production of the mortar composites, the possible interactions between the PCM and the other components, and the matrix durability [99,119]. For this latter evaluation, the resistance to water absorption, freeze/thaw tests, and biological colonization have been mainly used.

Only a few studies have dealt with the durability of PCMs in alkaline cementitious environments [138,141]. During hydration, cement particles dissolve, turning the pore solution into a caustic electrolyte. Indeed, the pore solution contains alkalis species (in particular, SO_4^{2-}, and Ca^{2+}) and, thus, typically exhibits a pH greater than 13 [142]. When micro-encapsulated PCMs are embedded in such caustic systems, chemical reactions between the pore solution and the capsule shell could take place, resulting in modifications and in a reduction in durability. The PCMs seem not to affect the corrosion behavior of reinforced concrete [49]; on the contrary, the incorporation of PCM could enhance the corrosion resistance by forming a protective film on the rebar surface, even if the concrete containing PCM develops a higher porosity compared to the un-modified concrete. The behavior and stability of PCM have been investigated in environments similar to those encountered in mortar and cement through immersion in alkaline solutions ($Ca(OH)_2$, NaOH) and in a saturated calcium sulfate solution. These tests highlighted an enthalpy reduction (around 25%) in the materials under study, mainly due to the rupture of the shell capsules with a consequent PCM leakage [119,138].

The low porosity due to the presence of PCM in mortars provides beneficial effects in terms of the composite's durability, due to the reduction in water absorption and hygroscopic capacity [84,89]. This effect has been ascribed to the ability of the PCM particles to obstruct and interrupt the capillary network [143]. A reduction in porosity can have a positive effect also towards the freeze-thaw resistance [84,122,144]; the incorporation of a PCM, on the other hand, generally results in higher losses of material during temperatures cycles below and above 0 °C.

The presence of additional components can decrease the bulk density, porosity, thermal conductivity and capillary water absorption of mortars, contributing to improve their durability, their resistance to biological colonization and to freeze-thaw action [77,80,85]. In addition, PCMs have been proposed for low-temperature applications as an alternative to de-icing salts, since they can improve the resistance against freeze-thaw cycles [93,145–147] and to reduce cracks formation in concrete. PCMs can reduce the time and depth of freezing in concrete by at least 10% [148].

6. Economic and Environmental Evaluation of Mortar with PCMs

Some considerations can be done on the overall costs of mortars containing PCM and on their applications in constructions. Starting from the assumption that using PCMs in buildings should lead to a reduction in energy consumption, thus, to an economic saving, there are different aspects to take into account. Usually, inorganic PCMs are cheaper than organic ones; among all the techniques used to incorporate PCMs in building materials, the micro- and macro-encapsulations are costly compared to the other methods (direct incorporation, direct impregnation). The shape- or form-stabilization techniques are promising, but still expensive, unless waste materials are used to produce PCM composite systems.

The incorporation of PCMs in mortars, improving the thermal comfort in the buildings, reduces the need for cooling and heating conditioning, leading to economic advantages. On the basis of numerical or experimental simulations, the energy costs, related to the capability of the mortars containing PCMs to decrease the heating and cooling needs (by increasing the minimum temperatures and decreasing the maximum temperatures), have been estimated to decrease by 10–50% [74,107], depending on the simulated season.

Very few studies report the assessment of applications of mortars with PCM using Life Cycle Analysis (LCA) and/or Life Cycle Cost Analysis (LCCA), used to evaluate the life cycle environmental impact and cost effects, respectively. Even though these techniques are already well developed and standardized, to the best of the authors' knowledge, there are no national or international standards available to test thermal energy storage products [149,150].

In general, PCMs in buildings can decrease the energy consumption, even though this may not imply an effective reduction in the global environmental impact throughout the lifetime of the structure [151]. Indeed, in most cases, the application of PCMs does not seem economically viable because of the high initial investment cost [149,152,153].

The environmental impact of PCMs could be greater than the conventional construction materials, depending on the type of PCM and the climate [105,154,155]. The benefits of the PCM increase in sites where the weather conditions are similar all year long [156]. Salt hydrates can be compensated in 25 years of use (while alkanes in 61 years) [150]. However, at the end of their useful life, most PCMs can be recycled; organic PCMs are biodegradable, and inorganic ones are innocuous. However, in order to minimize the overall environmental impact, the use of long-lasting PCMs, the extension of the useful life of the building, as well as the development of new PCMs with very low environmental impact, are suggested [156]. The use of PCM in heating and cooling systems in combination with conventional systems can further reduce the environmental impact.

7. Outlook for Future Works

Future research works on mortar/concrete containing PCMs are still needed to refine the methods of preparation, mitigate the strength reduction and overcome the durability issues. Possibly, different sustainable, green PCMs/support couples should be identified along with eco-efficient and low costly methods for the production of mortar/concrete containing PCMs. Very few studies have investigated the utility of geopolymers, even though this kind of mortar is eco-friendly and offers economic advantages, since industrial by-products and mine wastes can be used for their production.

Further studies should also be focused on economic and environmental assessments due to the presence of PCMs in building materials, since the current state-of-the-art about these topics is still lacking.

Experimental studies conducted in the field would be useful to document the performance and potential of the PCM in a real context. On the basis of these tests in real full-scale buildings, numerical models to assess the advantages and disadvantages of structures with or without PCMs (in advance) could be elaborated. Moreover, it would be desirable to develop simple but certified simulation tools that easily replicate the service conditions of PCM integrated into mortar or concrete in order to assess, using few tests, both the performance and durability of these materials.

8. Conclusions

In this paper, the use of PCMs in building materials was reviewed, especially for passive building systems, with the aim to outline their advantages in terms of thermal effects and thermal efficiency. In the first part of the paper, the main characteristics of PCMs were reported, describing the different and possible ways to introduce them in building materials. In the second part, a review of recently published experimental is presented, focusing on examples of PCMs introduced in mortars or concrete. Through this literature survey, the used materials and the supports, the preparation procedures (i.e., the different methods used to incorporate PCMs in building materials), the tests performed to analyze the final product, and the main results obtained in these studies are presented and discussed.

Thus, the following conclusions can be drawn.

- As far as the current world energy consumption is concerned, it is important to find alternative ways of saving energy and preserving the environment.
- PCMs are generally considered efficient materials that can improve thermal comfort in a building.
- The selection of the appropriate PCM for a specific construction material and/or a definite application must start from its properties (thermo-physical, chemical, functional, environmental and economic).

■ Among all the available methods to incorporate PCMs in building materials, the micro-encapsulation is the most used due to its advantages; however, in the last years, the form-stabilization method has gained popularity thanks to its low costs of production. Moreover, the latter is a promising technique due to the possibility to employ waste materials as a support for PCMs.

■ Concretes and mortars are considered suitable construction materials to incorporate PCMs since they are largely present in building constructions; furthermore, mortars can be applied in a building even after its construction.

■ The most widespread PCMs in building materials are organic in nature and have a melting temperature in a range between 20° and 40 °C.

■ The performance of micro-encapsulated PCM and those of pure PCM have been scarcely compared so far.

■ Chemical properties, thermal properties, and thermal stability are the main properties analyzed for PCMs. On the other hand, mortars/concretes with the addition of PCMs have been mainly studied in terms of their morphology, mechanical properties and thermal conductivity.

Author Contributions: Conceptualization, M.F.; analysis of literature, A.S. and M.L.; data curation, A.S. and M.L.; writing—original draft preparation, A.S.; writing—review and editing, M.L. and M.F.; supervision, M.F.

Funding: This research received no external funding.

Conflicts of Interest: The authors declare no conflict of interest.

References

1. Jeon, J.; Lee, J.-H.; Seo, J.; Jeong, S.-G.; Kim, S. Application of PCM thermal energy storage system to reduce building energy consumption. *J. Therm. Anal. Calorim.* **2013**, *111*, 279–288. [CrossRef]
2. Dean, B.; Dulac, J.; Petrichenko, K.; Graham, P. *Towards Zero-Emission Efficient and Resilient Buildings*; Global Status Report; Global Alliance for Buildings and Construction (GABC): Kongens Lyngby, Denmark, 2016.
3. Soares, N.; Costa, J.J.; Gaspar, A.R.; Santos, P. Review of passive PCM latent heat thermal energy storage systems towards buildings' energy efficiency. *Energy Build.* **2013**, *59*, 82–103. [CrossRef]
4. Lu, S.; Li, Y.; Kong, X.; Pang, B.; Chen, Y.; Zheng, S.; Sun, L. A Review of PCM Energy Storage Technology Used in Buildings for the Global Warming Solution. In *Energy Solutions to Combat Global Warming*; Zhang, X., Dincer, I., Eds.; Springer International Publishing: Cham, Switzerland, 2017; pp. 611–644. ISBN 978-3-319-26950-4.
5. Kalnæs, S.E.; Jelle, B.P. Phase change materials and products for building applications: A state-of-the-art review and future research opportunities. *Energy Build.* **2015**, *94*, 150–176. [CrossRef]
6. Janarthanan, B.; Sagadevan, S. Thermal energy storage using phase change materials and their applications: A review. *Int. J. ChemTech Res.* **2015**, *8*, 205–256.
7. Pomianowski, M.; Heiselberg, P.; Zhang, Y. Review of thermal energy storage technologies based on PCM application in buildings. *Energy Build.* **2013**, *67*, 56–69. [CrossRef]
8. Cui, Y.Q.; Riffat, S. Review on Phase Change Materials for Building Applications. *Appl. Mech. Mater.* **2011**, *71–78*, 1958–1962. [CrossRef]
9. Madessa, H.B. A Review of the Performance of Buildings Integrated with Phase Change Material: Opportunities for Application in Cold Climate. *Energy Procedia* **2014**, *62*, 318–328. [CrossRef]
10. Nkwetta, D.N.; Haghighat, F. Thermal energy storage with phase change material—A state-of-the art review. *Sustain. Cities Soc.* **2014**, *10*, 87–100. [CrossRef]
11. Telkes, M. Thermal storage for solar heating and cooling. In Proceedings of the Workshop on Solar Energy Storage Subsystems for the Heating and Cooling of Buildings, Charlottesville, VA, USA, 16–18 April 1975; pp. 17–23.
12. Lane, G.A. *Solar Heat Storage: Latent Heat Materials. Volume II. Technology*; CRC Press, Inc.: Boca Raton, FL, USA, 1986.
13. Akeiber, H.; Nejat, P.; Majid, M.Z.A.; Wahid, M.A.; Jomehzadeh, F.; Zeynali Famileh, I.; Calautit, J.K.; Hughes, B.R.; Zaki, S.A. A review on phase change material (PCM) for sustainable passive cooling in building envelopes. *Renew. Sustain. Energy Rev.* **2016**, *60*, 1470–1497. [CrossRef]

14. Rao, V.V.; Parameshwaran, R.; Ram, V.V. PCM-mortar based construction materials for energy efficient buildings: A review on research trends. *Energy Build.* **2018**, *158*, 95–122. [CrossRef]

15. Abhat, A. Low temperature latent heat thermal energy storage: Heat storage materials. *Sol. Energy* **1983**, *30*, 313–332. [CrossRef]

16. Cabeza, L.F.; Castell, A.; Barreneche, C.; de Gracia, A.; Fernández, A.I. Materials used as PCM in thermal energy storage in buildings: A review. *Renew. Sustain. Energy Rev.* **2011**, *15*, 1675–1695. [CrossRef]

17. Zhou, D.; Zhao, C.Y.; Tian, Y. Review on thermal energy storage with phase change materials (PCMs) in building applications. *Appl. Energy* **2012**, *92*, 593–605. [CrossRef]

18. Rathod, M.K.; Banerjee, J. Thermal stability of phase change materials used in latent heat energy storage systems: A review. *Renew. Sustain. Energy Rev.* **2013**, *18*, 246–258. [CrossRef]

19. Memon, S.A. Phase change materials integrated in building walls: A state of the art review. *Renew. Sustain. Energy Rev.* **2014**, *31*, 870–906. [CrossRef]

20. Pasupathy, A.; Velraj, R.; Seeniraj, R.V. Phase change material-based building architecture for thermal management in residential and commercial establishments. *Renew. Sustain. Energy Rev.* **2008**, *12*, 39–64. [CrossRef]

21. Socaciu, L.G. Thermal energy storage with phase change material. *Leonardo Electron. J. Pract. Technol.* **2012**, *11*, 75–98.

22. Whiffen, T.R.; Riffat, S.B. A review of PCM technology for thermal energy storage in the built environment: Part I. *Int. J. Low-Carbon Technol.* **2013**, *8*, 147–158. [CrossRef]

23. Kamali, S. Review of free cooling system using phase change material for building. *Energy Build.* **2014**, *80*, 131–136. [CrossRef]

24. Kenisarin, M.M. Thermophysical properties of some organic phase change materials for latent heat storage. A review. *Sol. Energy* **2014**, *107*, 553–575. [CrossRef]

25. Su, W.; Darkwa, J.; Kokogiannakis, G. Review of solid–liquid phase change materials and their encapsulation technologies. *Renew. Sustain. Energy Rev.* **2015**, *48*, 373–391. [CrossRef]

26. Thambidurai, M.; Panchabikesan, K.; Ramalingam, V. Review on phase change material based free cooling of buildings—The way toward sustainability. *J. Energy Storage* **2015**, *4*, 74–88. [CrossRef]

27. Cui, Y.; Xie, J.; Liu, J.; Pan, S. Review of Phase Change Materials Integrated in Building Walls for Energy Saving. *Procedia Eng.* **2015**, *121*, 763–770. [CrossRef]

28. Yuan, Y.; Zhang, N.; Tao, W.; Cao, X.; He, Y. Fatty acids as phase change materials: A review. *Renew. Sustain. Energy Rev.* **2014**, *29*, 482–498. [CrossRef]

29. Veerakumar, C.; Sreekumar, A. Phase change material based cold thermal energy storage: Materials, techniques and applications—A review. *Int. J. Refrig.* **2016**, *67*, 271–289. [CrossRef]

30. Souayfane, F.; Fardoun, F.; Biwole, P.-H. Phase change materials (PCM) for cooling applications in buildings: A review. *Energy Build.* **2016**, *129*, 396–431. [CrossRef]

31. Pereira da Cunha, J.; Eames, P. Thermal energy storage for low and medium temperature applications using phase change materials—A review. *Appl. Energy* **2016**, *177*, 227–238. [CrossRef]

32. Khan, Z.; Khan, Z.; Ghafoor, A. A review of performance enhancement of PCM based latent heat storage system within the context of materials, thermal stability and compatibility. *Energy Convers. Manag.* **2016**, *115*, 132–158. [CrossRef]

33. Amaral, C.; Vicente, R.; Marques, P.A.A.P.; Barros-Timmons, A. Phase change materials and carbon nanostructures for thermal energy storage: A literature review. *Renew. Sustain. Energy Rev.* **2017**, *79*, 1212–1228. [CrossRef]

34. Cui, Y.; Xie, J.; Liu, J.; Wang, J.; Chen, S. A review on phase change material application in building. *Adv. Mech. Eng.* **2017**, *9*. [CrossRef]

35. Zhai, X.Q.; Wang, X.L.; Wang, T.; Wang, R.Z. A review on phase change cold storage in air-conditioning system: Materials and applications. *Renew. Sustain. Energy Rev.* **2013**, *22*, 108–120. [CrossRef]

36. Waqas, A.; Ud Din, Z. Phase change material (PCM) storage for free cooling of buildings—A review. *Renew. Sustain. Energy Rev.* **2013**, *18*, 607–625. [CrossRef]

37. Iten, M.; Liu, S.; Shukla, A. A review on the air-PCM-TES application for free cooling and heating in the buildings. *Renew. Sustain. Energy Rev.* **2016**, *61*, 175–186. [CrossRef]

38. Safari, A.; Saidur, R.; Sulaiman, F.A.; Xu, Y.; Dong, J. A review on supercooling of Phase Change Materials in thermal energy storage systems. *Renew. Sustain. Energy Rev.* **2017**, *70*, 905–919. [CrossRef]

39. Kasaeian, A.; Bahrami, L.; Pourfayaz, F.; Khodabandeh, E.; Yan, W.-M. Experimental studies on the applications of PCMs and nano-PCMs in buildings: A critical review. *Energy Build.* **2017**, *154*, 96–112. [CrossRef]

40. Sun, Y.; Wang, S.; Xiao, F.; Gao, D. Peak load shifting control using different cold thermal energy storage facilities in commercial buildings: A review. *Energy Convers. Manag.* **2013**, *71*, 101–114. [CrossRef]

41. Huang, X.; Alva, G.; Jia, Y.; Fang, G. Morphological characterization and applications of phase change materials in thermal energy storage: A review. *Renew. Sustain. Energy Rev.* **2017**, *72*, 128–145. [CrossRef]

42. Song, M.; Niu, F.; Mao, N.; Hu, Y.; Deng, S. Review on building energy performance improvement using phase change materials. *Energy Build.* **2018**, *158*, 776–793. [CrossRef]

43. Alqallaf, H.J.; Alawadhi, E.M. Concrete roof with cylindrical holes containing PCM to reduce the heat gain. *Energy Build.* **2013**, *61*, 73–80. [CrossRef]

44. Johra, H.; Heiselberg, P. Influence of internal thermal mass on the indoor thermal dynamics and integration of phase change materials in furniture for building energy storage: A review. *Renew. Sustain. Energy Rev.* **2017**, *69*, 19–32. [CrossRef]

45. Silva, T.; Vicente, R.; Rodrigues, F. Literature review on the use of phase change materials in glazing and shading solutions. *Renew. Sustain. Energy Rev.* **2016**, *53*, 515–535. [CrossRef]

46. Sharma, A.; Tyagi, V.V.; Chen, C.R.; Buddhi, D. Review on thermal energy storage with phase change materials and applications. *Renew. Sustain. Energy Rev.* **2009**, *13*, 318–345. [CrossRef]

47. Kusama, Y.; Ishidoya, Y. Thermal effects of a novel phase change material (PCM) plaster under different insulation and heating scenarios. *Energy Build.* **2017**, *141*, 226–237. [CrossRef]

48. Lin, Y.; Jia, Y.; Alva, G.; Fang, G. Review on thermal conductivity enhancement, thermal properties and applications of phase change materials in thermal energy storage. *Renew. Sustain. Energy Rev.* **2018**, *82*, 2730–2742. [CrossRef]

49. Cellat, K.; Tezcan, F.; Beyhan, B.; Kardaş, G.; Paksoy, H. A comparative study on corrosion behavior of rebar in concrete with fatty acid additive as phase change material. *Constr. Build. Mater.* **2017**, *143*, 490–500. [CrossRef]

50. Pons, O.; Aguado, A.; Fernández, A.I.; Cabeza, L.F.; Chimenos, J.M. Review of the use of phase change materials (PCMs) in buildings with reinforced concrete structures. *Mater. Constr.* **2014**, *64*. [CrossRef]

51. Hassan, A.; Shakeel Laghari, M.; Rashid, Y. Micro-Encapsulated Phase Change Materials: A Review of Encapsulation, Safety and Thermal Characteristics. *Sustainability* **2016**, *8*, 1046. [CrossRef]

52. Jamekhorshid, A.; Sadrameli, S.M.; Farid, M. A review of microencapsulation methods of phase change materials (PCMs) as a thermal energy storage (TES) medium. *Renew. Sustain. Energy Rev.* **2014**, *31*, 531–542. [CrossRef]

53. Liu, C.; Rao, Z.; Zhao, J.; Huo, Y.; Li, Y. Review on nanoencapsulated phase change materials: Preparation, characterization and heat transfer enhancement. *Nano Energy* **2015**, *13*, 814–826. [CrossRef]

54. Abokersh, M.H.; Osman, M.; El-Baz, O.; El-Morsi, M.; Sharaf, O. Review of the phase change material (PCM) usage for solar domestic water heating systems (SDWHS). *Int. J. Energy Res.* **2017**, *42*, 329–357. [CrossRef]

55. Khadiran, T.; Hussein, M.Z.; Zainal, Z.; Rusli, R. Encapsulation techniques for organic phase change materials as thermal energy storage medium: A review. *Sol. Energy Mater. Sol. Cells* **2015**, *143*, 78–98. [CrossRef]

56. Giro-Paloma, J.; Martínez, M.; Cabeza, L.F.; Fernández, A.I. Types, methods, techniques, and applications for microencapsulated phase change materials (MPCM): A review. *Renew. Sustain. Energy Rev.* **2016**, *53*, 1059–1075. [CrossRef]

57. Alva, G.; Lin, Y.; Liu, L.; Fang, G. Synthesis, characterization and applications of microencapsulated phase change materials in thermal energy storage: A review. *Energy Build.* **2017**, *144*, 276–294. [CrossRef]

58. Mohamed, S.A.; Al-Sulaiman, F.A.; Ibrahim, N.I.; Zahir, M.H.; Al-Ahmed, A.; Saidur, R.; Yılbaş, B.S.; Sahin, A.Z. A review on current status and challenges of inorganic phase change materials for thermal energy storage systems. *Renew. Sustain. Energy Rev.* **2017**, *70*, 1072–1089. [CrossRef]

59. Saffari, M.; de Gracia, A.; Ushak, S.; Cabeza, L.F. Passive cooling of buildings with phase change materials using whole-building energy simulation tools: A review. *Renew. Sustain. Energy Rev.* **2017**, *80*, 1239–1255. [CrossRef]

60. Riffat, S.; Mempouo, B.; Fang, W. Phase change material developments: A review. *Int. J. Ambient Energy* **2015**, *36*, 102–115. [CrossRef]

61. Fokaides, P.A.; Kylili, A.; Kalogirou, S.A. Phase change materials (PCMs) integrated into transparent building elements: A review. *Mater. Renew. Sustain. Energy* **2015**, *4*, 6. [CrossRef]

62. Kibria, M.A.; Anisur, M.R.; Mahfuz, M.H.; Saidur, R.; Metselaar, I.H.S.C. A review on thermophysical properties of nanoparticle dispersed phase change materials. *Energy Convers. Manag.* **2015**, *95*, 69–89. [CrossRef]

63. Fallahi, A.; Guldentops, G.; Tao, M.; Granados-Focil, S.; Van Dessel, S. Review on solid-solid phase change materials for thermal energy storage: Molecular structure and thermal properties. *Appl. Therm. Eng.* **2017**, *127*, 1427–1441. [CrossRef]

64. Frigione, M.; Lettieri, M.; Sarcinella, A.; de Aguiar, J.B. Mortars with Phase Change Materials (PCM) and Stone Waste to Improve Energy Efficiency in Buildings. In Proceedings of the International Congress on Polymers in Concrete (ICPIC 2018), Washington DC, USA, 29 April–1 May 2018; Taha, M.M.R., Ed.; Springer International Publishing: Cham, Switzerland, 2018; pp. 195–201.

65. Maldonado-Alameda, A.; Lacasta, A.M.; Giro-Paloma, J.; Chimenos, J.M.; Formosa, J. Physical, thermal and mechanical study of MPC formulated with LG-MgO incorporating Phase Change Materials as admixture. *IOP Conf. Ser. Mater. Sci. Eng.* **2017**, *251*, 012024. [CrossRef]

66. Zhang, P.; Xiao, X.; Ma, Z.W. A review of the composite phase change materials: Fabrication, characterization, mathematical modeling and application to performance enhancement. *Appl. Energy* **2016**, *165*, 472–510. [CrossRef]

67. Khodadadi, J.M.; Fan, L.; Babaei, H. Thermal conductivity enhancement of nanostructure-based colloidal suspensions utilized as phase change materials for thermal energy storage: A review. *Renew. Sustain. Energy Rev.* **2013**, *24*, 418–444. [CrossRef]

68. Li, M.; Shi, J. Review on micropore grade inorganic porous medium based form stable composite phase change materials: Preparation, performance improvement and effects on the properties of cement mortar. *Constr. Build. Mater.* **2019**, *194*, 287–310. [CrossRef]

69. Lv, P.; Liu, C.; Rao, Z. Review on clay mineral-based form-stable phase change materials: Preparation, characterization and applications. *Renew. Sustain. Energy Rev.* **2017**, *68*, 707–726. [CrossRef]

70. Aguayo, M.; Das, S.; Maroli, A.; Kabay, N.; Mertens, J.C.E.; Rajan, S.D.; Sant, G.; Chawla, N.; Neithalath, N. The influence of microencapsulated phase change material (PCM) characteristics on the microstructure and strength of cementitious composites: Experiments and finite element simulations. *Cem. Concr. Compos.* **2016**, *73*, 29–41. [CrossRef]

71. Coppola, L.; Coffetti, D.; Lorenzi, S. Cement-Based Renders Manufactured with Phase-Change Materials: Applications and Feasibility. *Adv. Mater. Sci. Eng.* **2016**, *2016*, 1–6. [CrossRef]

72. Lecompte, T.; Le Bideau, P.; Glouannec, P.; Nortershauser, D.; Le Masson, S. Mechanical and thermo-physical behaviour of concretes and mortars containing phase change material. *Energy Build.* **2015**, *94*, 52–60. [CrossRef]

73. Eddhahak, A.; Drissi, S.; Colin, J.; Caré, S.; Neji, J. Effect of phase change materials on the hydration reaction and kinetic of PCM-mortars. *J. Therm. Anal. Calorim.* **2014**, *117*, 537–545. [CrossRef]

74. Cunha, S.; Aguiar, J.B.; Tadeu, A. Thermal performance and cost analysis of mortars made with PCM and different binders. *Constr. Build. Mater.* **2016**, *122*, 637–648. [CrossRef]

75. Ventolà, L.; Vendrell, M.; Giraldez, P. Newly-designed traditional lime mortar with a phase change material as an additive. *Constr. Build. Mater.* **2013**, *47*, 1210–1216. [CrossRef]

76. Lucas, S.S.; Ferreira, V.M.; Barroso de Aguiar, J.L. Latent heat storage in PCM containing mortars—Study of microstructural modifications. *Energy Build.* **2013**, *66*, 724–731. [CrossRef]

77. Santos, T.; Nunes, L.; Faria, P. Production of eco-efficient earth-based plasters: Influence of composition on physical performance and bio-susceptibility. *J. Clean. Prod.* **2018**, *167*, 55–67. [CrossRef]

78. Cunha, S.; Aguiar, J.B.; Ferreira, V.; Tadeu, A. Influence of Adding Encapsulated Phase Change Materials in Aerial Lime Based Mortars. *Adv. Mater. Res.* **2013**, *687*, 255–261. [CrossRef]

79. Cunha, S.; Aguiar J., B.; Kheradmand, M.; Bragança, L.; Ferreira V., M. Thermal Mortars with Incorporation of PCM Microcapsules. *RBM* **2014**, *19*, 171. [CrossRef]

80. Cunha, S.; Aguiar, J.; Ferreira, V.; Tadeu, A. Mortars based in different binders with incorporation of phase-change materials: Physical and mechanical properties. *Eur. J. Environ. Civ. Eng.* **2015**, *19*, 1216–1233. [CrossRef]

81. Cunha, S.; Aguiar, J.; Pacheco-Torgal, F. Effect of temperature on mortars with incorporation of phase change materials. *Constr. Build. Mater.* **2015**, *98*, 89–101. [CrossRef]

82. Cunha, S.; Aguiar, J.B.; Ferreira, V.M.; Tadeu, A. Influence of the Type of Phase Change Materials Microcapsules on the Properties of Lime-Gypsum Thermal Mortars. *Adv. Eng. Mater.* **2014**, *16*, 433–441. [CrossRef]

83. Cunha, S.; Aguiar, J.; Ferreira, V.; Tadeu, A.; Garbacz, A. Mortars with Phase Change Materials—Part I: Physical and Mechanical Characterization. *Key Eng. Mater.* **2015**, *634*, 22–32. [CrossRef]

84. Cunha, S.; Aguiar, J.B.; Ferreira, V.; Tadeu, A.; Garbacz, A. Mortars with Phase Change Materials—Part II: Durability Evaluation. *Key Eng. Mater.* **2015**, *634*, 33–45. [CrossRef]

85. Cunha, S.; Aguiar, J.; Tadeu, A. Ranking procedure based on mechanical, durability and thermal behavior of mortars with incorporation of phase change materials. *Mater. Constr.* **2015**, *65*, 1–5. [CrossRef]

86. Cunha, S.; Aguiar, J.; Ferreira, V.; Tadeu, A. Argamassas com incorporação de Materiais de Mudança de Fase (PCM): Caracterização física, mecânica e durabilidade. *Matéria* **2015**, *20*, 245–261. [CrossRef]

87. Cunha, S.; Aguiar, J.; Ferreira, V. Mortars with Incorporation of Phase Change Materials for Thermal Rehabilitation. *Int. J. Archit. Herit.* **2017**, *11*, 339–348. [CrossRef]

88. Pavlík, Z.; Trník, A.; Ondruška, J.; Keppert, M.; Pavlíková, M.; Volfová, P.; Kaulich, V.; Černý, R. Apparent Thermal Properties of Phase-Change Materials: An Analysis Using Differential Scanning Calorimetry and Impulse Method. *Int. J. Thermophys.* **2013**, *34*, 851–864. [CrossRef]

89. Pavlík, Z.; Trník, A.; Keppert, M.; Pavlíková, M.; Žumár, J.; Černý, R. Experimental Investigation of the Properties of Lime-Based Plaster-Containing PCM for Enhancing the Heat-Storage Capacity of Building Envelopes. *Int. J. Thermophys.* **2014**, *35*, 767–782. [CrossRef]

90. Wang, Z.; Su, H.; Zhao, S.; Zhao, N. Influence of phase change material on mechanical and thermal properties of clay geopolymer mortar. *Constr. Build. Mater.* **2016**, *120*, 329–334. [CrossRef]

91. Shadnia, R.; Zhang, L.; Li, P. Experimental study of geopolymer mortar with incorporated PCM. *Constr. Build. Mater.* **2015**, *84*, 95–102. [CrossRef]

92. Kheradmand, M.; Pacheco Torgal, F.; Azenha, M. Thermal Performance of Fly Ash Geopolymeric Mortars Containing Phase Change Materials. In Proceedings of the International Congress on Polymers in Concrete (ICPIC 2018), Washington DC, USA, 29 April–1 May 2018; Taha, M.M.R., Ed.; Springer International Publishing: Cham, Switzerland, 2018; pp. 565–570.

93. Pilehvar, S.; Szczotok, A.M.; Rodríguez, J.F.; Valentini, L.; Lanzón, M.; Pamies, R.; Kjøniksen, A.-L. Effect of freeze-thaw cycles on the mechanical behavior of geopolymer concrete and Portland cement concrete containing micro-encapsulated phase change materials. *Constr. Build. Mater.* **2019**, *200*, 94–103. [CrossRef]

94. Kim, S.; Paek, S.; Jeong, S.-G.; Lee, J.-H.; Kim, S. Thermal performance enhancement of mortar mixed with octadecane/xGnP SSPCM to save building energy consumption. *Sol. Energy Mater. Sol. Cells* **2014**, *122*, 257–263. [CrossRef]

95. Vieira, J.; Senff, L.; Gonçalves, H.; Silva, L.; Ferreira, V.M.; Labrincha, J.A. Functionalization of mortars for controlling the indoor ambient of buildings. *Energy Build.* **2014**, *70*, 224–236. [CrossRef]

96. Wang, P.; Li, N.; Zhao, C.; Wu, L.; Han, G. A Phase Change Storage Material that May be Used in the Fire Resistance of Building Structure. *Procedia Eng.* **2014**, *71*, 261–264. [CrossRef]

97. Joulin, A.; Zalewski, L.; Lassue, S.; Naji, H. Experimental investigation of thermal characteristics of a mortar with or without a micro-encapsulated phase change material. *Appl. Therm. Eng.* **2014**, *66*, 171–180. [CrossRef]

98. Kim, S.; Chang, S.J.; Chung, O.; Jeong, S.-G.; Kim, S. Thermal characteristics of mortar containing hexadecane/xGnP SSPCM and energy storage behaviors of envelopes integrated with enhanced heat storage composites for energy efficient buildings. *Energy Build.* **2014**, *70*, 472–479. [CrossRef]

99. He, Y.; Zhang, X.; Zhang, Y.; Song, Q.; Liao, X. Utilization of lauric acid-myristic acid/expanded graphite phase change materials to improve thermal properties of cement mortar. *Energy Build.* **2016**, *133*, 547–558. [CrossRef]

100. Li, M.; Wu, Z.; Tan, J. Heat storage properties of the cement mortar incorporated with composite phase change material. *Appl. Energy* **2013**, *103*, 393–399. [CrossRef]

101. Zhang, Z.; Shi, G.; Wang, S.; Fang, X.; Liu, X. Thermal energy storage cement mortar containing n-octadecane/ expanded graphite composite phase change material. *Renew. Energy* **2013**, *50*, 670–675. [CrossRef]

102. Snoeck, D.; Priem, B.; Dubruel, P.; De Belie, N. Encapsulated Phase-Change Materials as additives in cementitious materials to promote thermal comfort in concrete constructions. *Mater. Struct.* **2016**, *49*, 225–239. [CrossRef]

103. Aguiar, J.; Cunha, S.; Kheradmand, M. Phase change materials: Contribute to sustainable construction. In Proceedings of the Congresso Luso-Brasileiro de Materiais de Construção Sustentáveis, Guimarães, Portugal, 5–7 March 2014; pp. 19–28.

104. Pielichowska, K.; Pielichowski, K. Phase change materials for thermal energy storage. *Prog. Mater. Sci.* **2014**, *65*, 67–123. [CrossRef]

105. Baldassarri, C.; Sala, S.; Caverzan, A.; Lamperti Tornaghi, M. Environmental and spatial assessment for the ecodesign of a cladding system with embedded Phase Change Materials. *Energy Build.* **2017**, *156*, 374–389. [CrossRef]

106. Yu, Y.Y.; Liu, J.S.; Xing, S.S.; Zuo, J.; He, X. Experimental Research of Cement Mortar With Incorporated Lauric Acid/Expanded Perlite Phase-Change Materials. *J. Test. Eval.* **2017**, *45*, 20160021. [CrossRef]

107. Shafie-khah, M.; Kheradmand, M.; Javadi, S.; Azenha, M.; de Aguiar, J.L.B.; Castro-Gomes, J.; Siano, P.; Catalão, J.P.S. Optimal behavior of responsive residential demand considering hybrid phase change materials. *Appl. Energy* **2016**, *163*, 81–92. [CrossRef]

108. Tittelein, P.; Gibout, S.; Franquet, E.; Johannes, K.; Zalewski, L.; Kuznik, F.; Dumas, J.-P.; Lassue, S.; Bédécarrats, J.-P.; David, D. Simulation of the thermal and energy behaviour of a composite material containing encapsulated-PCM: Influence of the thermodynamical modelling. *Appl. Energy* **2015**, *140*, 269–274. [CrossRef]

109. Tittelein, P.; Gibout, S.; Franquet, E.; Zalewski, L.; Defer, D. Identification of Thermal Properties and Thermodynamic Model for a Cement Mortar Containing PCM by Using Inverse Method. *Energy Procedia* **2015**, *78*, 1696–1701. [CrossRef]

110. Haurie, L.; Serrano, S.; Bosch, M.; Fernandez, A.I.; Cabeza, L.F. Single layer mortars with microencapsulated PCM: Study of physical and thermal properties, and fire behaviour. *Energy Build.* **2016**, *111*, 393–400. [CrossRef]

111. Franquet, E.; Gibout, S.; Tittelein, P.; Zalewski, L.; Dumas, J.-P. Experimental and theoretical analysis of a cement mortar containing microencapsulated PCM. *Appl. Therm. Eng.* **2014**, *73*, 32–40. [CrossRef]

112. Fabiani, C.; Pisello, L.A.; D'Alessandro, A.; Ubertini, F.; Cabeza, F.L.; Cotana, F. Effect of PCM on the Hydration Process of Cement-Based Mixtures: A Novel Thermo-Mechanical Investigation. *Materials* **2018**, *11*, 871. [CrossRef]

113. Jayalath, A.; San Nicolas, R.; Sofi, M.; Shanks, R.; Ngo, T.; Aye, L.; Mendis, P. Properties of cementitious mortar and concrete containing micro-encapsulated phase change materials. *Constr. Build. Mater.* **2016**, *120*, 408–417. [CrossRef]

114. Kheradmand, M.; de Aguiar, J.B.; Azenha, M. Estimation of the specific enthalpy-temperature functions for plastering mortars containing hybrid mixes of phase change materials. *Int. J. Energy Environ. Eng.* **2014**, *5*, 81. [CrossRef]

115. Kheradmand, M.; Azenha, M.; de Aguiar, J.L.B.; Krakowiak, K.J. Thermal behavior of cement based plastering mortar containing hybrid microencapsulated phase change materials. *Energy Build.* **2014**, *84*, 526–536. [CrossRef]

116. Kheradmand, M.; Azenha, M.; de Aguiar, J.L.B.; Castro-Gomes, J. Experimental and numerical studies of hybrid PCM embedded in plastering mortar for enhanced thermal behaviour of buildings. *Energy* **2016**, *94*, 250–261. [CrossRef]

117. Richardson, A.; Heniegal, A.; Tindall, J. Optimal performance characteristics of mortar incorporating phase change materials and silica fume. *J. Green Build.* **2017**, *12*, 59–78. [CrossRef]

118. Ricklefs, A.; Thiele, A.M.; Falzone, G.; Sant, G.; Pilon, L. Thermal conductivity of cementitious composites containing microencapsulated phase change materials. *Int. J. Heat Mass Transf.* **2017**, *104*, 71–82. [CrossRef]

119. Wei, Z.; Falzone, G.; Wang, B.; Thiele, A.; Puerta-Falla, G.; Pilon, L.; Neithalath, N.; Sant, G. The durability of cementitious composites containing microencapsulated phase change materials. *Cem. Concr. Compos.* **2017**, *81*, 66–76. [CrossRef]

120. Cui, H.; Liao, W.; Mi, X.; Lo, T.Y.; Chen, D. Study on functional and mechanical properties of cement mortar with graphite-modified microencapsulated phase-change materials. *Energy Build.* **2015**, *105*, 273–284. [CrossRef]

121. Thiele, A.M.; Wei, Z.; Falzone, G.; Young, B.A.; Neithalath, N.; Sant, G.; Pilon, L. Figure of merit for the thermal performance of cementitious composites containing phase change materials. *Cem. Concr. Compos.* **2016**, *65*, 214–226. [CrossRef]

122. Kheradmand, M.; Castro-Gomes, J.; Azenha, M.; Silva, P.D.; de Aguiar, J.L.B.; Zoorob, S.E. Assessing the feasibility of impregnating phase change materials in lightweight aggregate for development of thermal energy storage systems. *Constr. Build. Mater.* **2015**, *89*, 48–59. [CrossRef]

123. Sharifi, N.P.; Jafferji, H.; Reynolds, S.E.; Blanchard, M.G.; Sakulich, A.R. Application of lightweight aggregate and rice husk ash to incorporate phase change materials into cementitious materials. *J. Sustain. Cem.-Based Mater.* **2016**, *5*, 349–369. [CrossRef]

124. Ramakrishnan, S.; Wang, X.; Sanjayan, J.; Wilson, J. Thermal performance assessment of phase change material integrated cementitious composites in buildings: Experimental and numerical approach. *Appl. Energy* **2017**, *207*, 654–664. [CrossRef]

125. Ramakrishnan, S.; Wang, X.; Sanjayan, J.; Wilson, J. Thermal Energy Storage Enhancement of Lightweight Cement Mortars with the Application of Phase Change Materials. *Procedia Eng.* **2017**, *180*, 1170–1177. [CrossRef]

126. Sharifi, N.P.; Sakulich, A. Application of phase change materials to improve the thermal performance of cementitious material. *Energy Build.* **2015**, *103*, 83–95. [CrossRef]

127. Nepomuceno, M.C.S.; Silva, P.D. Experimental evaluation of cement mortars with phase change material incorporated via lightweight expanded clay aggregate. *Constr. Build. Mater.* **2014**, *63*, 89–96. [CrossRef]

128. Cunha, S.; Aguiar, J.; Ferreira, V.; Tadeu, A. Physical and Mechanical Properties of Cement Mortars with Direct Incorporation of Phase Change Material. In Proceedings of the International Congress on Polymers in Concrete, Washington DC, USA, 29 April–1 May 2018; Springer: Berlin, Germany, 2018; pp. 203–209.

129. Cui, H.; Tang, W.; Qin, Q.; Xing, F.; Liao, W.; Wen, H. Development of structural-functional integrated energy storage concrete with innovative macro-encapsulated PCM by hollow steel ball. *Appl. Energy* **2017**, *185*, 107–118. [CrossRef]

130. Qian, T.; Li, J. Octadecane/C-decorated diatomite composite phase change material with enhanced thermal conductivity as aggregate for developing structural-functional integrated cement for thermal energy storage. *Energy* **2018**, *142*, 234–249. [CrossRef]

131. Li, M.; Lin, Z.; Wu, L.; Wang, J.; Gong, N. Applications of graphite-enabled phase change material composites to improve thermal performance of cementitious materials. *IOP Conf. Ser. Mater. Sci. Eng.* **2017**, *264*, 012013. [CrossRef]

132. Liu, F.; Wang, J.; Qian, X. Integrating phase change materials into concrete through microencapsulation using cenospheres. *Cem. Concr. Compos.* **2017**, *80*, 317–325. [CrossRef]

133. Zhang, X.; Jin, W.; Lv, Y.; Zhang, H.; Zhou, W.; Ding, F. Preparation and characterization of mortar mixes containing organic acid/expanded vermiculite composite PCM. *Funct. Mater.* **2017**, *24*, 481–489. [CrossRef]

134. Li, T.; Yuan, Y.; Zhang, N. Thermal properties of phase change cement board with capric acid/expanded perlite form-stable phase change material. *Adv. Mech. Eng.* **2017**, *9*, 1–8. [CrossRef]

135. Tao, N.; Huang, H. Application of Phase Change Material (PCM) in Concrete for Thermal Energy Storage. In Proceedings of the International Congress on Polymers in Concrete (ICPIC 2018), Washington DC, USA, 29 April–1 May 2018; Taha, M.M.R., Ed.; Springer International Publishing: Cham, Switzerland, 2018; pp. 187–193.

136. Sobolčiak, P.; Abdelrazeq, H.; Özerkan, N.G.; Ouederni, M.; Nógellová, Z.; AlMaadeed, M.A.; Karkri, M.; Krupa, I. Heat transfer performance of paraffin wax based phase change materials applicable in building industry. *Appl. Therm. Eng.* **2016**, *107*, 1313–1323. [CrossRef]

137. Choi, W.-C.; Khil, B.-S.; Chae, Y.-S.; Liang, Q.-B.; Yun, H.-D. Feasibility of Using Phase Change Materials to Control the Heat of Hydration in Massive Concrete Structures. *Sci. World J.* **2014**, *2014*, 1–6. [CrossRef]

138. Fernandes, F.; Manari, S.; Aguayo, M.; Santos, K.; Oey, T.; Wei, Z.; Falzone, G.; Neithalath, N.; Sant, G. On the feasibility of using phase change materials (PCMs) to mitigate thermal cracking in cementitious materials. *Cem. Concr. Compos.* **2014**, *51*, 14–26. [CrossRef]

139. Meshgin, P.; Xi, Y.; Li, Y. Utilization of phase change materials and rubber particles to improve thermal and mechanical properties of mortar. *Constr. Build. Mater.* **2012**, *28*, 713–721. [CrossRef]

140. Hunger, M.; Entrop, A.G.; Mandilaras, I.; Brouwers, H.J.H.; Founti, M. The behavior of self-compacting concrete containing micro-encapsulated Phase Change Materials. *Cem. Concr. Compos.* **2009**, *31*, 731–743. [CrossRef]

141. Sarı, A. Thermal reliability test of some fatty acids as PCMs used for solar thermal latent heat storage applications. *Energy Convers. Manag.* **2003**, *44*, 2277–2287. [CrossRef]

142. Bullard, J.W.; Jennings, H.M.; Livingston, R.A.; Nonat, A.; Scherer, G.W.; Schweitzer, J.S.; Scrivener, K.L.; Thomas, J.J. Mechanisms of cement hydration. *Cem. Concr. Res.* **2011**, *41*, 1208–1223. [CrossRef]

143. Kheradmand, M.; Abdollahnejad, Z.; Pacheco-Torgal, F. Alkali-activated cement-based binder mortars containing phase change materials (PCMs): Mechanical properties and cost analysis. *Eur. J. Environ. Civ. Eng.* **2018**, 1–23. [CrossRef]

144. Drissi, S.; Ling, T.-C. Thermal and durability performances of mortar and concrete containing phase change materials. *IOP Conf. Ser. Mater. Sci. Eng.* **2018**, *431*, 062001. [CrossRef]

145. Nayak, S.; Krishnan, N.M.A.; Das, S. Microstructure-guided numerical simulation to evaluate the influence of phase change materials (PCMs) on the freeze-thaw response of concrete pavements. *Constr. Build. Mater.* **2019**, *201*, 246–256. [CrossRef]

146. Li, W.; Ling, C.; Jiang, Z.; Yu, Q. Evaluation of the potential use of form-stable phase change materials to improve the freeze-thaw resistance of concrete. *Constr. Build. Mater.* **2019**, *203*, 621–632. [CrossRef]

147. Yeon, J.H.; Kim, K.-K. Potential applications of phase change materials to mitigate freeze-thaw deteriorations in concrete pavement. *Constr. Build. Mater.* **2018**, *177*, 202–209. [CrossRef]

148. Esmaeeli, H.S.; Farnam, Y.; Haddock, J.E.; Zavattieri, P.D.; Weiss, W.J. Numerical analysis of the freeze-thaw performance of cementitious composites that contain phase change material (PCM). *Mater. Des.* **2018**, *145*, 74–87. [CrossRef]

149. Konstantinidou, C.A.; Lang, W.; Papadopoulos, A.M.; Santamouris, M. Life cycle and life cycle cost implications of integrated phase change materials in office buildings. *Int. J. Energy Res.* **2019**, *43*, 150–166. [CrossRef]

150. Kyriaki, E.; Konstantinidou, C.; Giama, E.; Papadopoulos, A.M. Life cycle analysis (LCA) and life cycle cost analysis (LCCA) of phase change materials (PCM) for thermal applications: A review. *Int. J. Energy Res.* **2018**, *42*, 3068–3077. [CrossRef]

151. Akeiber, H.J.; Wahid, M.A.; Hussen, H.M.; Mohammad, A.T. Review of development survey of phase change material models in building applications. *Sci. World J.* **2014**, *2014*, 391690. [CrossRef] [PubMed]

152. Rincón, L.; Castell, A.; Pérez, G.; Solé, C.; Boer, D.; Cabeza, L.F. Evaluation of the environmental impact of experimental buildings with different constructive systems using Material Flow Analysis and Life Cycle Assessment. *Appl. Energy* **2013**, *109*, 544–552. [CrossRef]

153. Cabeza, L.F.; Rincón, L.; Vilariño, V.; Pérez, G.; Castell, A. Life cycle assessment (LCA) and life cycle energy analysis (LCEA) of buildings and the building sector: A review. *Renew. Sustain. Energy Rev.* **2014**, *29*, 394–416. [CrossRef]

154. Aranda-Usón, A.; Ferreira, G.; López-Sabirón, A.M.; Mainar-Toledo, M.D.; Zabalza Bribián, I. Phase change material applications in buildings: An environmental assessment for some Spanish climate severities. *Sci. Total Environ.* **2013**, *444*, 16–25. [CrossRef] [PubMed]

155. Serrano, S.; Barreneche, C.; Rincón, L.; Boer, D.; Cabeza, L.F. Optimization of three new compositions of stabilized rammed earth incorporating PCM: Thermal properties characterization and LCA. *Constr. Build. Mater.* **2013**, *47*, 872–878. [CrossRef]

156. De Gracia, A.; Rincón, L.; Castell, A.; Jiménez, M.; Boer, D.; Medrano, M.; Cabeza, L.F. Life Cycle Assessment of the inclusion of phase change materials (PCM) in experimental buildings. *Energy Build.* **2010**, *42*, 1517–1523. [CrossRef]

MDPI

St. Alban-Anlage 66

4052 Basel

Switzerland

Tel. +41 61 683 77 34

Fax +41 61 302 89 18

www.mdpi.com

Materials Editorial Office

E-mail: materials@mdpi.com

www.mdpi.com/journal/materials

www.ingramcontent.com/pod-product-compliance
Lightning Source LLC
Chambersburg PA
CBHW051852210326
41597CB00033B/5870